D1754519

Energy Storage

Scrivener Publishing
100 Cummings Center, Suite 541J
Beverly, MA 01915-6106

Publishers at Scrivener
Martin Scrivener (martin@scrivenerpublishing.com)
Phillip Carmical (pcarmical@scrivenerpublishing.com)

Energy Storage

Edited by
Umakanta Sahoo

WILEY

This edition first published 2021 by John Wiley & Sons, Inc., 111 River Street, Hoboken, NJ 07030, USA and Scrivener Publishing LLC, 100 Cummings Center, Suite 541J, Beverly, MA 01915, USA
© 2021 Scrivener Publishing LLC
For more information about Scrivener publications please visit www.scrivenerpublishing.com.

All rights reserved. No part of this publication may be reproduced, stored in a retrieval system, or transmitted, in any form or by any means, electronic, mechanical, photocopying, recording, or otherwise, except as permitted by law. Advice on how to obtain permission to reuse material from this title is available at http://www.wiley.com/go/permissions.

Wiley Global Headquarters
111 River Street, Hoboken, NJ 07030, USA

For details of our global editorial offices, customer services, and more information about Wiley products visit us at www.wiley.com.

Limit of Liability/Disclaimer of Warranty
While the publisher and authors have used their best efforts in preparing this work, they make no representations or warranties with respect to the accuracy or completeness of the contents of this work and specifically disclaim all warranties, including without limitation any implied warranties of merchantability or fitness for a particular purpose. No warranty may be created or extended by sales representatives, written sales materials, or promotional statements for this work. The fact that an organization, website, or product is referred to in this work as a citation and/or potential source of further information does not mean that the publisher and authors endorse the information or services the organization, website, or product may provide or recommendations it may make. This work is sold with the understanding that the publisher is not engaged in rendering professional services. The advice and strategies contained herein may not be suitable for your situation. You should consult with a specialist where appropriate. Neither the publisher nor authors shall be liable for any loss of profit or any other commercial damages, including but not limited to special, incidental, consequential, or other damages. Further, readers should be aware that websites listed in this work may have changed or disappeared between when this work was written and when it is read.

Library of Congress Cataloging-in-Publication Data

ISBN 978-1-119-55551-3

Cover image: (Antenna Tower): Carmen Hauser | Dreamstime.com
Cover design by Kris Hackerott

Set in size of 11pt and Minion Pro by Manila Typesetting Company, Makati, Philippines

10 9 8 7 6 5 4 3 2 1

Contents

List of Contributors	xi
Preface	xiii

1 Thermal Energy Storage Systems for Concentrating Solar Power Plants   1
Dr. Pratibha Biswal

1.1	Introduction		2
1.2	Concentrating Solar Power (CSP) Technology		2
	1.2.1	CSP Receiver Concepts	4
		1.2.1.1 Parabolic Trough System	4
		1.2.1.2 Linear Fresnel Reflector Systems	5
		1.2.1.3 Central Receiver Plants	6
		1.2.1.4 Dish System	7
1.3	Thermal Energy Storage in CSP		7
	1.3.1	Active Two-Tank System	9
		1.3.1.1 Active Two-Tank Direct	9
	1.3.2	Active Single-Tank Thermocline	20
	1.3.3	Other TES Systems	21
		1.3.3.1 Packed-Bed Storage System	21
		1.3.3.2 Passive Thermal Storage System	22
	1.3.4	Types of Thermal Energy Storage (TES)	22
		1.3.4.1 Sensible Energy Storage	22
		1.3.4.2 Latent Heat Storage	24
		1.3.4.3 Thermochemical Energy Storage	25
1.4	Corrosion Problem in TES-CSP System		26
1.5	Conclusion		26
	References		27

2 Solar Thermal Power Plant with Thermal Energy Storage   31
Anil Kumar, Umakanta Sahoo and BK Jayasimha Rathod

2.1	Introduction	32

	2.2	Literature Review	39
		2.2.1 Power Installed Capacity of India	39
		2.2.2 Energy Storage Systems	40
		2.2.3 Thermal Storage Systems	40
	2.3	Energy Demand of World	44
	2.4	Experimental Set Up	48
		2.4.1 Description of Experimental Set Ups	49
	2.5	Experimental Data Analysis, Results and Discussions	55
		2.5.1 Performance of Reflector Round the Year (Experimental Set up I)	58
		2.5.1.1 Simulation Results	63
		2.5.1.2 Typical PID of a Solar Module from 'India One' Solar Power Plant	66
		2.5.1.3 Quantity of Steam to Turbine	67
	2.6	Experimental Data Analysis, Results and Discussions	69
	2.7	Conclusions	75
		Symbols	76
		Acknowledgement	77
		References	77
3	**Efficient Energy Storage Systems for Wind Power Application**		**81**
	Pradeep Kumar Sahu, Satyaranjan Jena and Umakanta Sahoo		
	3.1	Introduction	82
	3.2	Energy Storage Devices	84
		3.2.1 Electrical Energy Storage	84
		3.2.1.1 Superconducting Magnetic Energy Storage (SMES)	85
		3.2.1.2 Supercapacitors	86
		3.2.2 Mechanical Energy Storage	87
		3.2.2.1 Flywheel Energy Storage (FES)	87
		3.2.2.2 Pumped Hydroelectric Storage (PHS)	88
		3.2.2.3 Compressed Air Energy Storage	89
		3.2.3 Chemical Energy Storage	89
		3.2.3.1 Battery Storage System (BSS)	90
		3.2.3.2 Fuel Cells	90
		3.2.3.3 Solar Fuel	90
		3.2.4 Thermal Energy Storage	93
	3.3	Hybrid Energy Storage System (HESS)	93
	3.4	Power Converter Topologies for Hybrid Energy Storage	95
		3.4.1 Passive Topology	95
		3.4.2 Semi-Active Topology	97

		3.4.3	Active Topology	97
		3.4.4	Comparison of Different Topologies	98
	3.5	HESS Energy Management and Control		99
		3.5.1	HESS Control Schemes	99
			3.5.1.1 Classical Control Scheme	100
			3.5.1.2 Intelligent Control Schemes	102
		3.5.2	Comparison of Different Control Schemes	103
	3.6	Applications of the Storage Technologies in Wind Power		104
		3.6.1	Power Fluctuation Mitigation	104
		3.6.2	Low Voltage Ride Through (LVRT)	105
		3.6.3	Voltage Control Support	105
		3.6.4	Oscillation Damping	106
		3.6.5	Peak Shaving	106
		3.6.6	Spinning Reserve	107
		3.6.7	Time Shifting	108
		3.6.8	Transmission Line Curtailment	108
		3.6.9	Load Following	109
		3.6.10	Unit Commitment	110
	3.7	Conclusion		110
		References		112
4	**Advances in Electrochemical Energy Storage Device: Supercapacitor**			**119**
	Swagatika Kamila, Bikash Kumar Jena and Suddhasatwa Basu			
	4.1	Introduction		120
	4.2	Types of Energy Storage Devices		120
	4.3	Overview of Supercapacitor and Its Global Scenario		122
	4.4	Status of Supercapacitor in India		125
	4.5	Types of Supercapacitor According to the Energy Storage Mechanism		126
		4.5.1	Electrical Double-Layer Capacitor (EDLC)	126
		4.5.2	Pseudocapacitor	128
		4.5.3	Hybrid Supercapacitor	129
			4.5.3.1 Composite Supercapacitor	129
			4.5.3.2 Asymmetric Supercapacitor	130
			4.5.3.3 Battery Type	130
	4.6	Basic Components of Supercapacitor		130
		4.6.1	Current Collector	130
		4.6.2	Electrode Materials	131
			4.6.2.1 EDLC Materials	131

			4.6.2.2	Pseudocapacitive Materials	132
		4.6.3	Electrolytes		138
		4.6.4	Binders		138
		4.6.5	Separators		139
	4.7	Conclusion			140
		References			140

5 **Thermal Energy Storage Systems for Cooling and Heating Applications** — **149**
Pankaj Kalita, Debangsu Kashyap and Urbashi Bordoloi

	5.1	Introduction			150
	5.2	Classification of Storage Systems			151
	5.3	Sensible Heat Storage			151
		5.3.1	Water-Based Storage		153
		5.3.2	Packed Beds		156
		5.3.3	Aquifers		158
		5.3.4	Borehole		160
	5.4	Latent Heat Storage			163
		5.4.1	Enhancement Methods for Thermal Conductivity Enhancement		164
			5.4.1.1	Macro and Microencapsulation	165
			5.4.1.2	Addition of Fins	166
			5.4.1.3	Multiple PCM Technology	167
			5.4.1.4	Immersion Through Material Pores	167
	5.5	Thermochemical Heat Storage			168
		5.5.1	Absorption Cycle		172
		5.5.2	Adsorption Cycles		173
		5.5.3	Chemical Reaction		174
	5.6	Application of Thermal Energy Storage Systems			176
		5.6.1	Absorption Refrigeration System		176
		5.6.2	Solar Pumps Application in Space Cooling/Heating		177
		5.6.3	Solar Pond Integrated Packed-Bed TES System for Space Heating		178
		5.6.4	Solar FPC		179
		5.6.5	Solar PV/T		181
		5.6.6	Solar Air Heater		183
	5.7	Design Problems			184
	5.8	Conclusion			196
		References			196

6 Optimistic Technological Approaches for Sustainable Energy Storage Devices/Materials 201
Benjamin Raj, Arya Das, Suddhasatwa Basu and Mamata Mohapatra
- 6.1 Introduction 202
- 6.2 Advancements in Supercapacitor Technology 202
  - 6.2.1 The Current Global Supercapacitor Market 205
  - 6.2.2 Challenges: From Lab to Market 207
  - 6.2.3 Current Trends and Opportunities 209
  - 6.2.4 Composites and Novel Architectures 209
  - 6.2.5 Microsupercapacitors 210
  - 6.2.6 Hybrid Supercapacitors 211
  - 6.2.7 Flexible, Wearable and Smart Supercapacitors 211
- 6.3 Advancements in Battery Technology 212
  - 6.3.1 Challenges 213
  - 6.3.2 Nickel-Cadmium Batteries 213
  - 6.3.3 Nickel-Metal Hydride Batteries 214
  - 6.3.4 Lead Storage Battery 214
  - 6.3.5 Sodium Sulphur Battery 215
  - 6.3.6 Flow Batteries 217
  - 6.3.7 Lithium Ion Batteries (LIBs) 218
- 6.4 Conclusion and Outlook 221
- References 222

7 Electro-Chemical Battery Energy Storage Systems - A Comprehensive Overview 229
Nikhil P G and G Sivaramakrishnan
- 7.1 Introduction 229
- 7.2 Electro-Chemical Storage Devices 231
  - 7.2.1 Definition and Types 231
  - 7.2.2 Energy Storage Landscape and Benefits of Electro-Chemical Storage 235
  - 7.2.3 Drivers and Barriers in Implementation of Energy Storage Systems 240
- 7.3 Design and Performance Parameters for Electro-Chemical Storage 240
  - 7.3.1 Design Basis for Large Storage Application 240
- 7.4 Case Study From Industry 243
- 7.5 Best Practices in Battery Maintenance 245
- 7.6 End of Life Cycle of Batteries 247

x CONTENTS

		7.6.1	Major Recyclable Products from the Process	248
		7.6.2	Disposal Measures	248
	7.7	India Energy Storage Mission		249
	7.8	Conclusion		251
		References		251

8 Simulation of Charging and Discharging a Thermal Energy Storage System Involving Phase Change Material 253
S. Sanyal, A. Borgohain and S.P. Gupta

	8.1	Introduction		253
	8.2	Design of Latent Heat Storage (LHS) System		256
		8.2.1	Identification of Suitable PCM	256
		8.2.2	Design of Heat Exchanger	260
		8.2.3	Performance Evaluation	261
	8.3	Analysis of Phase Change Systems		261
	8.4	Simulation		263
		8.4.1	Equations Involved	263
		8.4.2	Modelling	265
		8.4.3	Transient Analysis	269
	8.5	Results and Discussion		269
		8.5.1	Scalability of Mesh	269
		8.5.2	Melting	270
		8.5.3	Solidification	271
		8.5.4	Performance	273
	8.6	Conclusion		274
		Acknowledgement		274
		Abbreviation		275
		References		275

Index   277

List of Contributors

Pratibha Biswal, Indian Institute of Petroleum and Energy, Visakhapatnam, Andhra Pradesh, India

Anil Kumar, Ministry of New and Renewable Energy, CGO Complex, Lodhi Road, New Delhi, India

Umakanta Sahoo, National Institute of Solar Energy, Gwal Pahari, Haryana, India

BK Jayasimha Rathod, CEO, India One Solar Plant, Brahmakumaris, Shantivan, Abu Road, Rajasthan, India

Pradeep Kumar Sahu, School of Electrical Engineering, KIIT, Deemed to be University, Bhubaneswar, India

Satyaranjan Jena, School of Electrical Engineering, KIIT, Deemed to be University, Bhubaneswar, India

Suddhasatwa Basu, Materials Chemistry Department, CSIR-Institute of Minerals and Materials Technology, Bhubaneswar, 751013, India and Academy of Scientific & Innovative Research (AcSIR), Ghaziabad-201002, India

Swagatika Kamila, Materials Chemistry Department, CSIR-Institute of Minerals and Materials Technology, Bhubaneswar, 751013, India and Academy of Scientific & Innovative Research (AcSIR), Ghaziabad-201002, India

Bikash Kumar Jena, Materials Chemistry Department, CSIR-Institute of Minerals and Materials Technology, Bhubaneswar, 751013, India and Academy of Scientific & Innovative Research (AcSIR), Ghaziabad-201002, India

Pankaj Kalita, Centre for Energy, Indian Institute of Technology Guwahati, Guwahati-781039, Assam

Debangsu Kashyap, Centre for Energy, Indian Institute of Technology Guwahati, Guwahati-781039, Assam

Urbashi Bordoloi, Centre for Rural Technology, Indian Institute of Technology Guwahati, Guwahati-781039, Assam

Benjamin Raj, CSIR-Institute of Minerals and Materials Technology, Bhubaneswar-751013, Odisha, India

Arya Das, CSIR-Institute of Minerals and Materials Technology, Bhubaneswar-751013, Odisha, India

Mamata Mohapatra, CSIR-Institute of Minerals and Materials Technology, Bhubaneswar-751013, Odisha, India

Nikhil P G, National Institute of Solar Energy, Gwal Pahari, Haryana, India

G. Sivaramakrishnan, Chartered Engineer, Institution of Engineers (India) Kolkata, West Bengal 700020, India

S. Sanyal, Central University of Jharkhand, Ranchi, Jharkhand 835205, India

A. Borgohain, Bhabha Atomic Research Centre, Mumbai, Maharashtra 400094, India

S.P. Gupta, Bhabha Atomic Research Centre, Mumbai, Maharashtra 400094, India

Preface

The energy scene in the world is a complex picture of a variety of energy sources being used to meet the growing energy needs. However, there is a gap in the demand and supply position. It is recognized that decentralized generation based on the various renewable energy technologies can, to some extent, help in meeting the growing energy needs. The renewable energy landscape in the world, during the last few years, has witnessed tremendous changes in the policy framework with accelerated and ambitious plans to increase the contribution of renewable energy such as solar, wind, bio-power, etc. Energy storage is one of the options for continuous operation from renewable energy sources; they complement each other and offer several benefits over a stand-alone system. Renewable energy can enhance the capacity utilization of energy storage systems in a true sense and lead to greater security of continuous electricity supply and other applications. Each chapter is written by renowned experts in the field of energy storage. The book covers various types of energy storage, i.e., thermal energy storage and chemical energy storage as well as other emerging technologies.

This book on energy storage aims to provide a platform for researchers, academicians, industry professionals, consultants and designers to present the state-of-the-art developments and challenges in the field of energy storage and in particular renewable energy for sustainability and scalability.

1

Thermal Energy Storage Systems for Concentrating Solar Power Plants

Dr. Pratibha Biswal

Indian Institute of Petroleum and Energy, Visakhapatnam, Andhra Pradesh, India

Abstract

This chapter presents the relevance and integration of TES for CSP technologies. A TES system consists of the storage material, heat transfer equipment, and storage tank. The TES material stores the thermal energy either in the form of sensible heat, latent heat and thermochemical energy via chemical reactions. There are several requirements that must be considered to ensure optimal storage dynamics and longevity in a TES. These requirements are analysed and discussed. A broad spectrum of storage technologies, materials and methods are explored for the selection of suitable TES for CSP technologies. Materials for heat transfer fluid and material for energy storage that are generally used in TES are presented. Various limitations and problems of TES systems, such as high temperature corrosion with their proposed solutions, as well as successful implementations are reported. Further, storage media and storage type selection for CSPs based on their stability, material characterization and compatibility of materials are explained. Various available CSP technologies such as parabolic trough collector technology, linear Fresnel collector technology, solar tower technology, Stirling dish technology, etc., are discussed in detail and compared. Factors to be considered at different hierarchical levels for each CSP technologies with TES are explained.

Keywords: Solar thermal, thermal energy storage, parabolic trough collector, solar tower, stirling dish technology

Email: biswal.pratibha9@gmail.com

1.1 Introduction

According to the International Energy Agency (IEA) in 2019, global energy demand will rise by 1.3% each year to 2040. Non-renewable energy takes up a major percentage of the global energy sector. The awareness and use of renewable energy is one of the ways to meet sustainable energy goals and to alleviate the associated environmental problems including carbon emission. Also, the development on renewable energy is significantly slow-paced as observed in the last few decades. The development of renewable energy is seen only in very few countries. The most popular renewable energy sources currently under consideration are wind energy, solar energy, tidal energy, geothermal energy, hydro energy, etc. Due to many benefits including ready availability, the larger proportion of interest is taken by solar energy among various types of renewable energy sources (Regin et al., 2008). However, the momentum of renewable energy technologies is not as significant as is needed with the expansion of the global economy and growth of population.

In the solar energy field, two major breakthrough technologies that have attracted significant attention in many countries are Solar Photovoltaic (PV) power generation and Concentrated Solar Power (CSP) plants. In solar PV plants, the solar energy is directly converted to electricity by using solar cells (Muhammad and Arshad, 2020; Jain et al., 2020; Khajepour and Ameri, 2020; Ahmed et al., 2020; Zhang et al., 2013). The solar energy to electrical energy conversion efficiency of a commercial grade PV is observed to be less than that of CSP systems and power dispatch is not possible in PV-based solar power production. Other than larger energy efficiency, one of the major advantages of CSP is its ability to provide electrical power at nighttime and during cloudy hours. This can be made possible by incorporating Thermal Energy Storage (TES) system. The capacity to dispatch power is more in a CSP system than that of a PV-based solar power system.

1.2 Concentrating Solar Power (CSP) Technology

Solar energy is the most viable and abundant renewable energy source. The Concentrating Solar Power (CSP) technology is promising especially for countries having an abundance of solar resources. Implementation of CSP technology can secure the energy supply and reduce carbon footprint, resulting in achieving sustainable development goals. A Concentrating Solar Power (CSP) system includes a concentrator (to concentrate solar radiation), a receiver (converts solar radiation to thermal energy) and a power block (with the turbine to convert thermal energy to electrical energy). A CSP system receives and concentrates sunlight followed by converting solar radiation to thermal energy (Shouman and Khattab, 2015; Tian and Zhao, 2013; Cavallaro, 2009; Barlev *et al.*, 2011; Desai *et al.*, 2014; Islam *et al.*, 2018). The thermal energy is then carried by a fluid called Heat Transfer Fluid (HTF) to the power block for power generation.

The CSP concept materialized on an industry scale in the 1980s in California where nine separate Solar Electric Generating Systems (SEGS) based on parabolic trough receiver, totalling 354 MWe of installed capacity were constructed. These systems used oil as the HTF involving parabolic trough receivers based on steam turbines for power generation. As observed from the data provided by NREL, the growth of solar power plants based on CSP concepts has been led predominantly by Spain followed by the United States. New and ongoing CSP projects are also being developed in other countries as shown in Figure 1.1. As per the data provided by NREL,

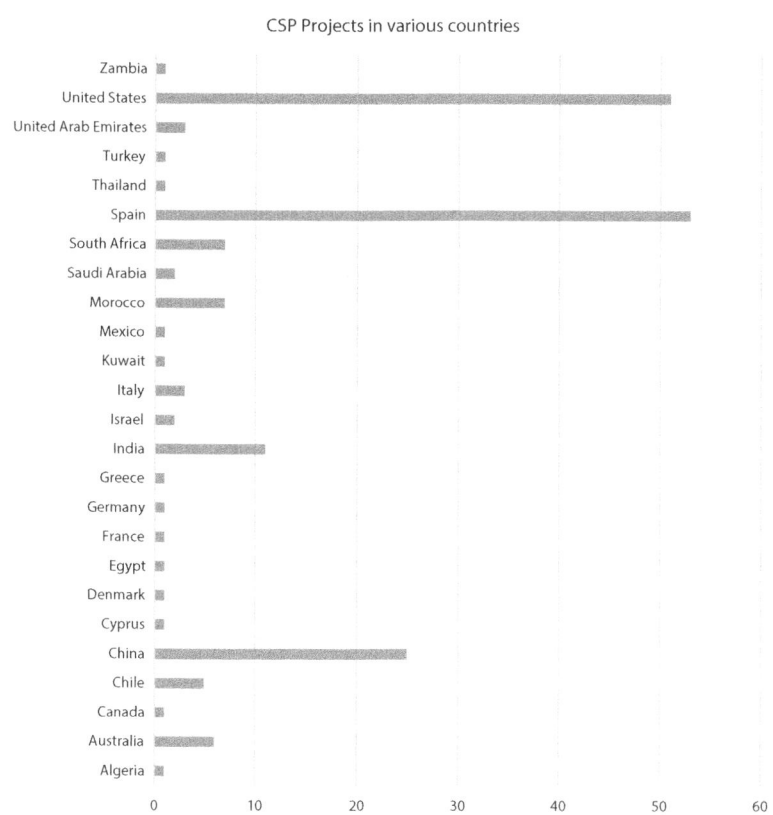

Figure 1.1 Number of CSP projects in various countries (source: NREL).

4 ENERGY STORAGE

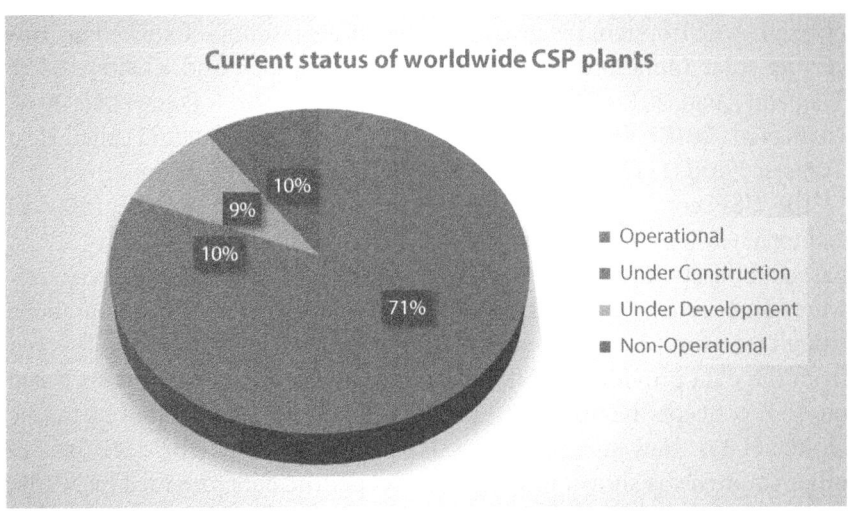

Figure 1.2 Current status of worldwide CSP plants (data source: NREL).

currently 188 worldwide CSP-based power plants are documented, out of which 71% are operational [see Figure 1.2].

1.2.1 CSP Receiver Concepts

There are two ways by which the solar radiation is concentrated from the solar panel: line focusing and point focusing systems. In the line focusing system, solar collectors concentrate the radiation along a focal line. In the point focusing system, solar radiation is concentrated on a single focal point. A single axis tracking system is needed to follow the sun during the day for the line-focusing systems. On the other hand, two axis tracking systems are needed for the mirrors in point focusing systems. Based on the focusing concepts and receiver geometry, currently four design concepts are used, such as parabolic trough system, linear Fresnel system, power towers and parabolic dish systems. A summary of these receiver concepts is illustrated in Figure 1.3 and briefly explained below.

1.2.1.1 Parabolic Trough System

In parabolic trough systems, solar radiation is focused to a receiver tube located at the focal point using mirrored parabolic troughs to focus. The receiver tube carries the HTF which can be heated up to a temperature of

Energy Storage in Concentrating Solar Power 5

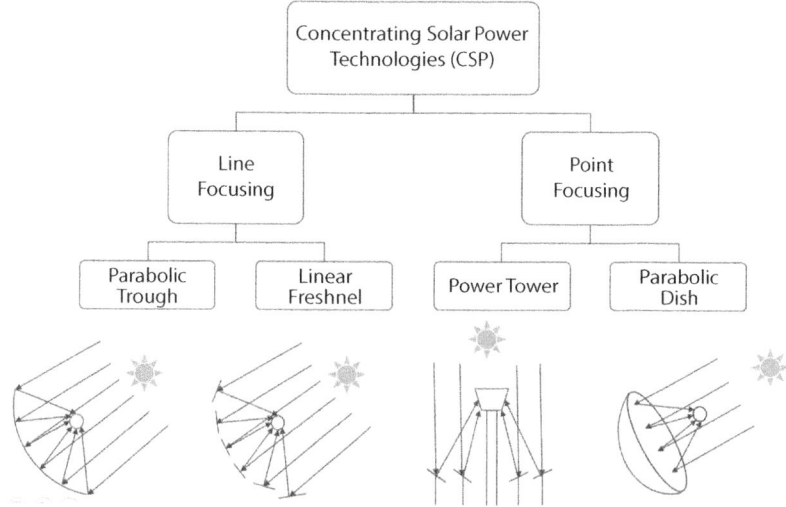

Figure 1.3 CSP receiver concepts.

390°C. The HTF is then pumped through a heat exchanger where superheated steam at high pressure is produced. Further, the steam is fed into a steam turbine connected to a generator to produce electricity. Synthetic oil, direct steam and molten salt are a few examples of HTF used in parabolic trough systems. The parabolic shaped reflectors implement a line-focusing system.

Parabolic trough systems are one of the promising technologies and most preferred configurations of CSP in the solar energy market. Most of today's commercial CSP plants are based on this concept, making up 65% (101 out of 188) of total CSP plants [Figure 1.4]. There are a good number of research works found in the literature on the parabolic trough system (Herrmann and Kearney, 2002; Herrmann *et al.*, 2004; Tamme *et al.*, 2004; Llorente *et al.*, 2011; Kolb, 2011).

1.2.1.2 Linear Fresnel Reflector Systems

Linear Fresnel reflector systems use a field of narrow long mirrors. The mirrors are rotated independently to concentrate solar radiation on a stationary receiver tube. The mirrors are either flat or curved that track the sun and focus solar radiation [Figure 1.3]. The fixed absorber tube allows for easier Direct Steam Generation (DSG). Due to direct DSG, the operating temperature for such systems can be larger. This system also works with line focusing concept.

6 ENERGY STORAGE

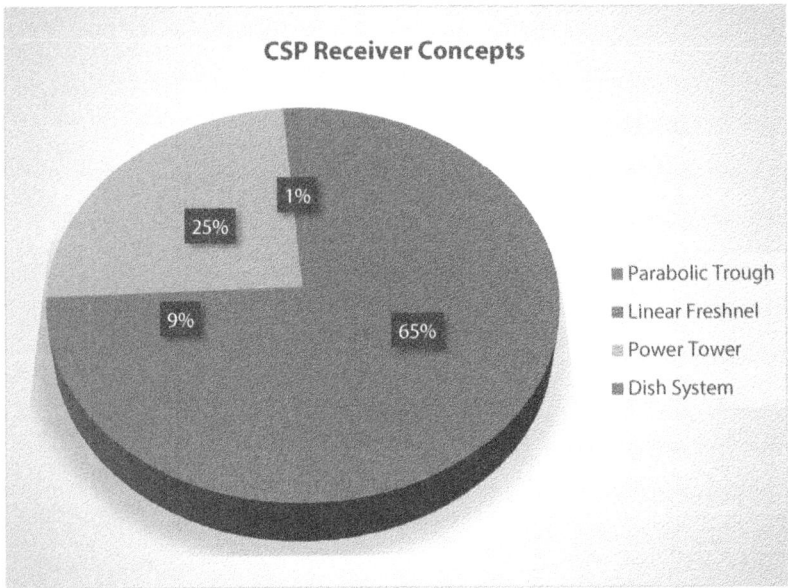

Figure 1.4 CSP receiver concepts as used worldwide.

One of the major advantages of such systems is the simple design of the reflector leading to lesser capital cost. Overall, Linear Fresnel CSP receiver system offers lowest start-up and maintenance cost. In addition, cheap and simple flat mirrors allow high reflectors density per square meter. However, linear Fresnel receivers have the lowest solar-to-electrical efficiency compared to other receivers due to high optical loss. Now 9% of all CSP plants opt for such collector concept [Figure 1.4]. As observed from the literature, there is a good amount of fundamental and academic research work going on to enhance the performance of the CSP system with linear Fresnel receiver (Desai and Bandyopadhyay, 2015; Mills *et al.*, 2000; Morin *et al.*, 2012; Xie *et al.*, 2011).

1.2.1.3 Central Receiver Plants

A circular array of large mirrors with sun-tracking motion (called heliostats) reflect direct sunlight onto an absorber system in central receiver. Such systems are also called power tower as the receiver is mounted on the top of a tower [Figure 1.3]. This is a point focusing system and the two axis tracking allows a higher concentration ratio and temperature. The receiver collects radiation as thermal energy and a HTF carries the energy from the receiver to the power block. Central receiver configuration allows high plant size due to its design. Currently, 25% of the installed CSP configurations are central receiver plants [Figure 1.4]. Research works on central receiver systems are reviewed and presented by Behar *et al.* (2013), Ho and Iverson (2013), and Avila-Marin (2011).

1.2.1.4 Dish System

In dish systems, dish-shaped parabolic mirrors are used as reflectors to concentrate and focus the solar radiation on a receiver [Figure 1.3]. The receiver is generally mounted above the dish. The receiver absorbs the thermal energy and transfers it to the Stirling engine or steam engine. Dish systems are also point focusing concept like solar towers. There were two thermal power plants with dish systems constructed [Figure 1.4]. In addition, very few works on dish systems are available in the literature (Zapata, 2015; Poullikkas *et al.*, 2010; Affandi *et al.*, 2015). Currently, dish systems have not been operational since the 1990s after the deployment of other configurations. The disadvantages of such systems are that they are expensive with a lower collection temperature.

1.3 Thermal Energy Storage in CSP

The intermittent and uncertain nature of solar energy are critical issues which cause a mismatch between source availability and energy demand. As a consequence, this results in ripples in the deployment and market penetrability of CSP technologies. Use of a backup system that uses non-renewable energy is one of the solutions of such problems. In addition, another cleaner solution to this problem is use of a Thermal Energy Storage (TES) system. The TES system can store energy when solar radiation is abundant. The stored energy can be released during periods without the solar radiation. This results in uninterrupted electricity generation during nighttime or on a cloudy day. Since the last few years, TES has been an explicit and significant component of CSP plant with the ability to drive the turbine almost continuously. Thus, electricity can be generated continuously at night and during cloudy hours when sunlight is not available. In addition, CSP with TES provides utility-scale and dispatchable electricity to the power grid. Also, TES allows CSP plants to generate electricity during evening hours when electricity is highly valued, offering better cost effectiveness. The integrated TES system can also result in shorter start-up time of the absorber systems. The TES system can be installed after the receiver system as illustrated in the block diagrams of Figure 1.5.

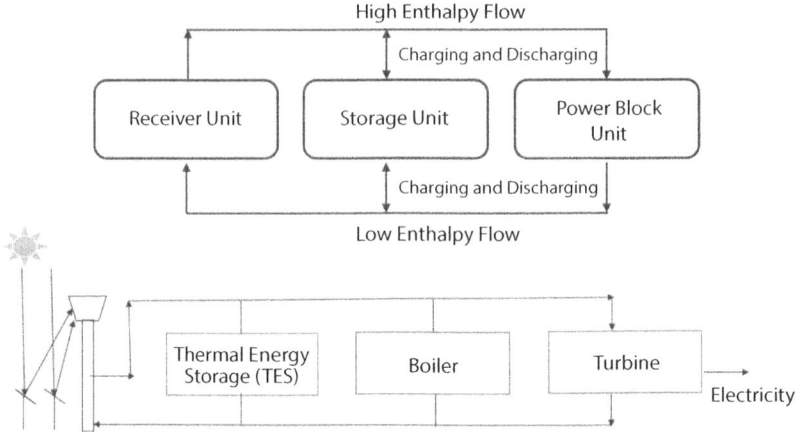

Figure 1.5 Integration of thermal energy storage systems to a CSP plant.

The possibility of integrating the TES system is one of the most distinct advantages of the solar energy field with CSP over other renewable energy fields. There are many storage technologies being developed that can be integrated into CSP plants. To accelerate the development of solar energy power generation in terms of economy and operation, an efficient and cost-effective TES system is important. Many earlier research works have studied the importance of TES on CSP and this results in an energy-efficient power plant (Guo *et al.*, 2018; John *et al.*, 2013; Prasad and Muthukumar, 2013; Adinberg, 2011; Powell and Edgar, 2012; Tamme and Laing, 2004; Tian and Zhao, 2013).

Thermal energy storage concepts for CSP plants can be classified as active or passive systems. In an active TES system, the thermal energy storage material itself circulates through a heat exchanger. Active systems can be divided into direct or indirect TES systems. In the commercial CSP plants, only active TES systems are used. Commercial active direct thermal energy storage systems are molten salt systems and steam accumulators. A second medium is used for storing the thermal energy in an active indirect system. A heat exchanger is used to transfer thermal energy from HTF to the second medium.

Two different materials for heat storage and circulation are used for passive storage systems which are dual medium storage systems. The HTF passes through the thermal storage system only for charging and discharging a TES material. The TES material can store energy in terms of sensible or latent heat. Passive TES systems are still under major research and they

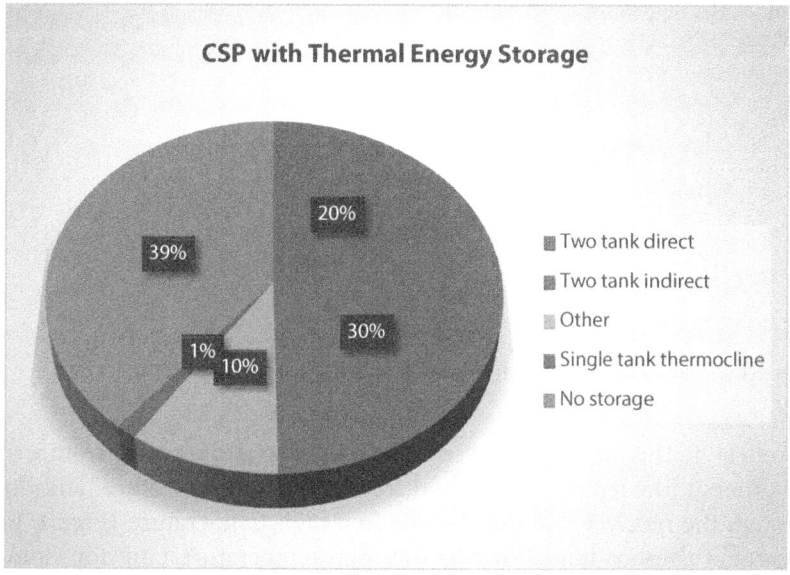

Figure 1.6 Reported integration of thermal energy storage systems in worldwide CSP plants.

have not yet reached the commercial CSP plant level. A brief discussion about different TES systems [Figure 1.6] as incorporated in commercial CSP follows below.

1.3.1 Active Two-Tank System

In an Active two-tank system, one tank holds hot HTF and another tank holds cold HTF. The hot HTF and cold HTF do not interact with each other. Major advantages of such systems are there will be no interaction between the fluids. Thus, heat loss is less and temperatures of the fluids in the tanks are almost uniform with time and position. However, two separate tanks of equal volumes are needed for such systems and tanks of the walls are subjected to daily cycles of molten salts. Thermal CSP plants with two-tank-based molten salt system are well documented by Kelly and Kearney (2004) and Herrmann *et al.* (2004). This concept can be used as direct or indirect storage systems.

1.3.1.1 *Active Two-Tank Direct*

In an active two-tank direct system, solar radiation is converted to thermal energy and the thermal energy is stored in the same fluid which was used

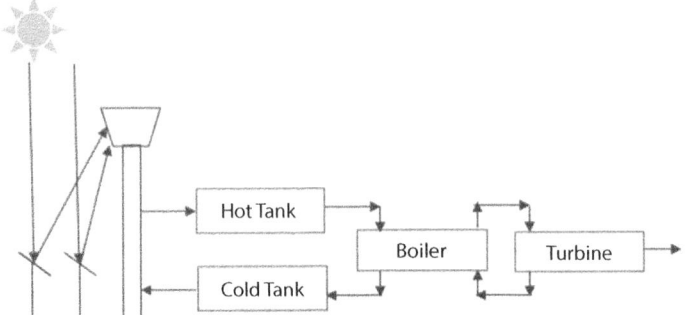

Figure 1.7 Active two-tank direct thermal energy storage.

to collect it. The fluid is stored in two tanks, one at high temperature and the other at low temperature. Fluid from the low-temperature tank flows through the receiver and gets heated to a high temperature [Figure 1.7]. Thereafter the hot fluid flows to the high-temperature tank for storage. Fluid from the high-temperature tank flows through a heat exchanger. In the heat exchanger the fluid loses heat to generates steam which is further used for electricity production. The cold fluid exits the heat exchanger and returns to the low-temperature tank to continue the cycle.

Concentrating solar power (CSP) projects that use parabolic trough systems such as Archimede, ASE Demo Plant, Chabei, Gansu Akesai, Solar Electric Generating Station I (SEGS I) use active two-tank direct thermal energy system [see Table 1.1]. Dacheng Dunhuang Molten Salt Fresnel project used two-tank direct storage with a storage capacity of 13 hours and Molten Salt as the heat transfer fluid. Such storage systems have attracted a great deal of attention from power tower receiver concept. Power tower projects such as Atacama-1, Aurora Solar Energy Project, Copiapó, Crescent Dunes Solar Energy Project (Tonopah) implement two-tank direct storage system. Currently, 20% of the commercial plants use two-tank direct TES system for energy storage, and the summary of all plants is listed in Tables 1.1, 1.2 and 1.3.

1.3.1.1.1 Active Two-Tank Indirect

In two-tank indirect systems, different fluids are used as the HTF and storage material. In cases when HTF is expensive or HTF is not well suited to be used as storage materials, such a system can be implemented. The storage material from the low-temperature storage tank flows through an extra heat exchanger. The fluid is heated by the high-temperature HTF in the

Table 1.1 CSP projects with parabolic trough receiver concepts integrated with thermal energy storage systems (data collected from NREL).

Parabolic trough CSP projects	Thermal energy storage (TES)	Thermal energy storage capacity	Storage description and material
Airlight Energy Ait-Baha Pilot Plant	Packed bed	5 hour(s)	Packed-bed of rocks
Andasol-1 (AS-1)	2-tank indirect	7.5 hour(s)	Molten salt 60% sodium nitrate and 40% potassium nitrate.
Andasol-2 (AS-2)	2-tank indirect	7.5 hour(s)	Molten salt 60% sodium nitrate and 40% potassium nitrate.
Andasol-3 (AS-3)	2-tank indirect	7.5 hours	Molten salts
Archimede	2-tank direct	8 hour(s)	Molten salt 60% sodium nitrate and 40% potassium nitrate.
Arcosol 50 (Valle 1)	2-tank indirect	7.5 hour(s)	Molten salt 60% sodium nitrate and 40% potassium nitrate.
Arenales	2-tank indirect	7 hours	Molten salt 60% sodium nitrate and 40% potassium nitrate.
ASE Demo Plant	2-tank direct	4.27 MWh-t	Molten salt
Ashalim (Negev)	2-tank indirect	4.5 hours	Molten salts

(Continued)

Table 1.1 CSP projects with parabolic trough receiver concepts integrated with thermal energy storage systems (data collected from NREL). (*Continued*)

Parabolic trough CSP projects	Thermal energy storage (TES)	Thermal energy storage capacity	Storage description and material
Aste 1A	2-tank indirect	8 Hours	Molten salts 60% sodium nitrate and 40% potassium nitrate
Aste 1B	2-tank indirect	8 Hours	Molten salts 60% sodium nitrate and 40% potassium nitrate
Astexol II	2-tank indirect	8 Hours	Molten salt 60% sodium nitrate and 40% potassium nitrate
Bokpoort	2-tank indirect	9.3 hours	Molten salt
Casablanca	2-tank indirect	7.5 hours	Molten salt 60% sodium nitrate and 40% potassium nitrate.
Chabei 64MW Molten Salt Parabolic Trough project	2-tank direct	16 hours	Molten salt
Delingha 50MW Thermal Oil Parabolic Trough project	2-tank indirect	9 hours	Molten salt
DEWA CSP Trough Project	2-tank indirect	15 hours	Molten salt

(*Continued*)

Table 1.1 CSP projects with parabolic trough receiver concepts integrated with thermal energy storage systems (data collected from NREL). (*Continued*)

Parabolic trough CSP projects	Thermal energy storage (TES)	Thermal energy storage capacity	Storage description and material
Diwakar	2-tank indirect	4 hours	Molten Salt
Extresol-1 (EX-1)	2-tank indirect	7.5 hour(s)	Molten salt 60% sodium nitrate and 40% potassium nitrate.
Extresol-2 (EX-2)	2-tank indirect	7.5 hour(s)	Molten salt 60% sodium nitrate and 40% potassium nitrate.
Extresol-3 (EX-3)	2-tank indirect	7.5 hour(s)	Molten salt 60% sodium nitrate and 40% potassium nitrate.
Gansu Akesai 50MW Molten Salt Trough project	2-tank direct	15 hours	Molten salt
Gujarat Solar One	2-tank indirect	9 hours	Molten salt
Gulang 100MW Thermal Oil Parabolic Trough project	2-tank indirect	7 hours	Molten salt
Ilanga I	2-tank indirect	5 hours	Molten salt
Kathu Solar Park	2-tank indirect	4.5 hours	Molten salt
KaXu Solar One	2-tank indirect	2.5 hours	Molten salts
KVK Energy Solar Project	2-tank indirect	4 hours	Molten salt

(*Continued*)

Table 1.1 CSP projects with parabolic trough receiver concepts integrated with thermal energy storage systems (data collected from NREL). (*Continued*)

Parabolic trough CSP projects	Thermal energy storage (TES)	Thermal energy storage capacity	Storage description and material
La Africana	2-tank indirect	7.5 hours	Molten salt 60% sodium nitrate and 40% potassium nitrate.
La Dehesa	2-tank indirect	7.5 hour(s)	Molten salt 60% sodium nitrate and 40% potassium nitrate.
La Florida	2-tank indirect	7.5 hour(s)	Molten salt 60% sodium nitrate and 40% potassium nitrate.
Manchasol-1 (MS-1)	2-tank indirect	7.5 hour(s)	Molten salt 60% sodium nitrate and 40% potassium nitrate.
Manchasol-2 (MS-2)	2-tank indirect	7.5 hour(s)	Molten salt 60% sodium nitrate and 40% potassium nitrate.
NOOR I	2-tank indirect	3 hours	Molten salt
NOOR II	2-tank indirect	7 hours	Molten salt
Rayspower Yumen 50MW Thermal Oil Trough project	2-tank indirect	7 hours	Molten salt

(*Continued*)

Table 1.1 CSP projects with parabolic trough receiver concepts integrated with thermal energy storage systems (data collected from NREL). (*Continued*)

Parabolic trough CSP projects	Thermal energy storage (TES)	Thermal energy storage capacity	Storage description and material
Shagaya CSP Project	2-tank indirect	9 hours	Molten salt
Solana Generating Station (Solana)	2-tank indirect	6 hours	Molten salt
Solar Electric Generating Station I (SEGS I)	2-tank direct	3 hour(s)	Molten salt
Termesol 50 (Valle 2)	2-tank indirect	7.5 hour(s)	Molten salt 60% sodium nitrate and 40% potassium nitrate.
Termosol 1	2-tank indirect	9 hours	Molten salt 60% sodium nitrate and 40% potassium nitrate.
Termosol 2	2-tank indirect	9 hours	Molten salt 60% sodium nitrate and 40% potassium nitrate.
Urat Royal Tech 100MW Thermal Oil Parabolic Trough project	2-tank indirect	10 hours	Molten Salt
Xina Solar One	2-tank indirect	5 hours	Molten salt
Yumen 50MW Thermal Oil Trough CSP project	2-tank indirect	7 hours	Molten salt

Table 1.2 CSP projects with linear Fresnel receiver concepts integrated with thermal energy storage systems (Data collected from NREL).

Linear fresnel CSP projects	Thermal energy storage (TES)	Thermal energy storage capacity	Storage description and material
Dacheng Dunhuang 50MW Molten Salt Fresnel project	2-tank direct	13 hours	Molten salt
eCare Solar Thermal Project	Other	2 hours	Steam drum
eLLO Solar Thermal Project (Llo)	Other	4 hours	Steam drum
IRESEN 1 MWe CSP-ORC pilot project	Other	20 minutes	Buffer
Puerto Errado 2 Thermosolar Power Plant (PE2)	Single-tank thermocline	0.5 Hours	Ruth's tank
Urat 50MW Fresnel CSP project	2-tank indirect	6 hours	Molten salt
Zhangbei 50MW DSG Fresnel CSP project	Other	14 hours	Solid state formulated concrete
Zhangjiakou 50MW CSG Fresnel project	Other	14 hours	Solid state formulated concrete

Table 1.3 CSP projects with central receiver (power tower) concepts integrated with thermal energy storage systems.

Power tower CSP projects	Thermal energy storage (TES)	Thermal energy storage capacity	Storage description and material
Atacama-1	2-tank direct	17.5 hours	Molten salt
Aurora Solar Energy Project	2-tank direct	8 hours	Molten salt
Copiapó	2-tank direct	14 hours	Molten salt
Crescent Dunes Solar Energy Project (Tonopah)	2-tank direct	10 hours	Molten salt
Dahan Power Plant	Other	1 hour	Saturated steam/oil
DEWA CSP Tower Project	2-tank direct	15 hours	Molten salt
Gemasolar Thermosolar Plant (Gemasolar)	2-tank direct	15 hour(s)	Molten salt
Golden Tower 100MW Molten Salt project	2-tank direct	8 hours	Molten salt
Golmud	2-tank direct	15 hours	Molten salt
Greenway CSP Mersin Tower Plant	Other	4 MW/h	Molten salt
Hami 50 MW CSP Project	2-tank direct	8 hours	Molten salt

(*Continued*)

Table 1.3 CSP projects with central receiver (power tower) concepts integrated with thermal energy storage systems. (*Continued*)

Power tower CSP projects	Thermal energy storage (TES)	Thermal energy storage capacity	Storage description and material
Huanghe Qinghai Delingha 135 MW DSG Tower CSP Project	2-tank indirect	3.7 hours	Molten salt
Jemalong Solar Thermal Station	2-tank direct	3 hours	Liquid sodium
Jülich Solar Tower	Other	1.5 hours	Ceramic heat sink
Khi Solar One	Other	2 hours	Saturated steam
Likana Solar Energy Project	2-tank direct	13 hours	Molten salt
MINOS	2-tank indirect	5 hours	Molten salt 60% sodium nitrate and 40% potassium nitrate
NOOR III	2-tank direct	7 hours	Molten salt
Qinghai Gonghe 50 MW CSP Plant	2-tank direct	6 hours	Molten salt
Redstone Solar Thermal Power Plant	2-tank direct	12 hours	Molten salt
Shangyi 50MW DSG Tower CSP project	2-tank indirect	4 hours	Molten salt

(*Continued*)

Table 1.3 CSP projects with central receiver (power tower) concepts integrated with thermal energy storage systems. (*Continued*)

Power tower CSP projects	Thermal energy storage (TES)	Thermal energy storage capacity	Storage description and material
Shouhang Dunhuang 10 MW Phase I	2-tank direct	15 hours	Molten salt
Shouhang Dunhuang 100 MW Phase II	2-tank direct	11 hours	Molten salt
SUPCON Delingha 10 MW Tower	2-tank direct	2 hours	Molten salt
SUPCON Delingha 50 MW Tower	2-tank direct	7 hours	Molten salt
Tamarugal Solar Energy Project	2-tank direct	13 hours	Molten salt
Yumen 100MW Molten Salt Tower CSP project	2-tank direct	10 hours	Molten salt
Yumen 50MW Molten Salt Tower CSP project	2-tank direct	6 hours	Molten salt

heat exchanger. The high-temperature storage material then returns back to the high-temperature storage tank. The HTF exits the heat exchanger at a low temperature returning to the receiver system to absorb heat again to continue the cycle. As observed in a two-tank direct system, storage material from the high-temperature tank is further used to generate steam. In contrast to the direct system, an extra heat exchanger is needed in an indirect system, making this system costlier.

Gujarat Solar One (Technology Parabolic Trough), a commercial plant developed by Cargo Solar Power, operates at a temperature range of 293°C-393°C and 9 hours storage capacity involving molten salt. Extresol-1, a commercial plant in Spain, developed by ACS/Cobra Group, uses storage system of capacity 7.5 hour(s). The storage material is a combination of various materials with 60% sodium nitrate and 40% potassium nitrate. Similarly, a large number of concentrating solar power (CSP) projects with all receiver concepts (except for dish system) use an active two-tank indirect storage system. The summary is listed in Tables 1.1, 1.2 and 1.3.

1.3.2 Active Single-Tank Thermocline

In a thermocline system, a single tank is used where warm fluid is above colder fluid. These kinds of systems are less expensive than two-tank systems [Figure 1.8]. The heat loss is more in such systems as the hot and cold

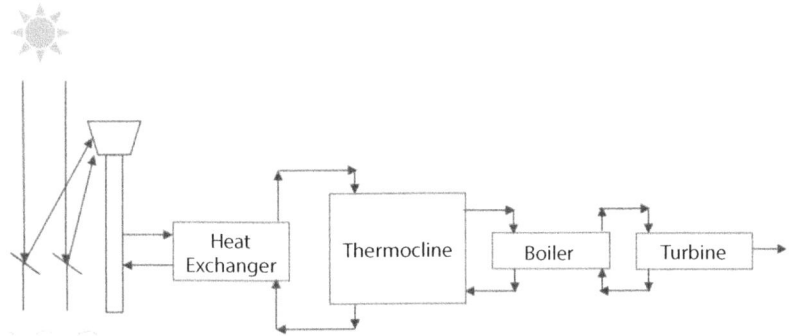

Figure 1.8 Active single tank thermocline thermal energy storage.

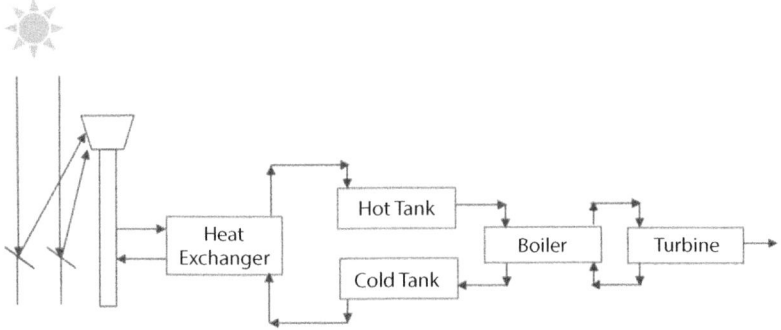

Figure 1.9 Active two-tank indirect thermal energy storage.

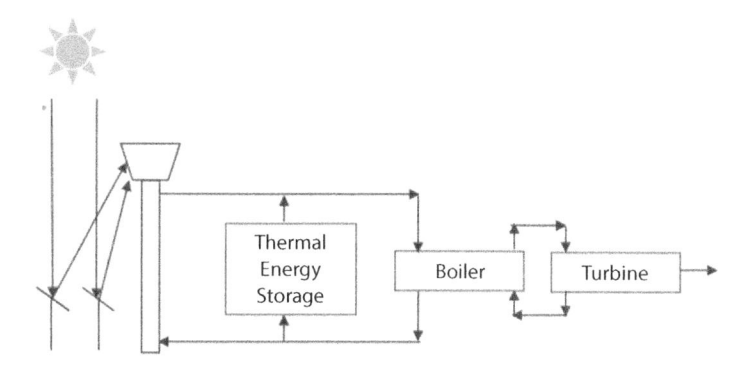

Figure 1.10 Passive thermal energy storage.

fluids are in direct contact. The hot and cold temperature regions are separated by a temperature gradient resulting in a thermocline. The density difference in the fluid thermally stratify the fluid in the tank. Buoyancy effects create thermal stratification of the fluid within the tank, which helps to stabilize and maintain the thermocline. Van Lew *et al.* (2011), Bayón and Rojas (2014), and Biencinto *et al.* (2014) carried out theoretical and experimental work on thermocline energy storage system for CSP plants.

High-temperature HTF flows into the top of the thermocline and leaves the bottom at low temperature. The thermocline moves downward and adds thermal energy to the system for storage. The thermocline moves upward and removes thermal energy from the system to generate steam and electricity if the flow is reversed. Puerto Errado 2 Thermosolar [see Table 1.2] operated by Novatec Solar España uses linear Fresnel reflector system currently operational in Calasparra, Spain, and has implemented a single-tank thermocline thermal energy storage system. This plant is operated at a temperature range of 140ºC-270ºC. Other plants also use this type of thermal systems and are listed in Tables 1.1, 1.2 and 1.3. Active two tank indirect thermal energy storage is also given in Figure 1.9 and passive type thermal energy storage in built with solar tower technology is given in Figure 1.10.

1.3.3 Other TES Systems

1.3.3.1 Packed-Bed Storage System

Other types of thermal energy storage system include packed-bed and passive system. Only power plants use a packed-bed system as the storage method is Airlight Energy Ait-Baha Pilot Plant. This plant uses Parabolic Trough Technology at temperature range of 270ºC-570ºC with 5 hours storage capacity.

1.3.3.2 Passive Thermal Storage System

In the passive type TES systems, thermal storage material is fixed and it does not flow, which is in contrast to the active system. The thermal storage material is used only to store thermal energy which can be transferred to and from the heat transfer fluid via thermal charging and discharging. A passive-type thermal storage system can be a solid material (example: concrete), fluid (example: water) or phase change material (example: PCM). In such systems, the heat transfer fluid transfer energy to the thermal storage material where the material stores energy which can be further transferred to the heat transfer fluid. Such systems have not been integrated in solar thermal power plants till date. The works on passive thermal energy storage system are on the laboratory and fundamental level. A good number of research works can be found in literature on the passive thermal energy storage system.

1.3.4 Types of Thermal Energy Storage (TES)

There are three types of TES mechanisms that can be applied to CSP and other applications: sensible energy storage, latent energy storage and thermochemical energy storage. An overview of these technical concepts and their states of development are presented below.

1.3.4.1 Sensible Energy Storage

The sensible thermal storage system stores thermal energy with increases in the temperature of the TES material. The principle of the sensible thermal storage system is simple and it has been widely applied in CSP as well as other applications. The TES material undergoes temperature change during energy storage and release. The physical and chemical changes of storage material is not observed. Sensible thermal energy storage method is simple and inexpensive. One of the disadvantages of sensible thermal energy storage material is low thermal conductivity. This results in lesser energy storage and release capacity. Also, the sensible heat transfer materials have low energy storage density. This further leads to large sizes of the storage devices.

The amount of energy stored in the material (Q) can be calculated as

$$Q = mC_p \Delta T$$

Where m is the mass of the material, C_p is the specific heat of the material at constant pressure and ΔT is the temperature difference.

Common sensible storage materials include water, steam, synthetic oil, molten salt, gravel, etc. [see Table 1.4]. As seen from Tables 1.1, 1.2 and 1.3, molten salt is widely used for sensible energy storage. Sensible thermal storage systems are mainly seen for low-temperature applications.

Research on sensible thermal storage is comparatively mature and has been developed to a commercially exploitative level. As the density of sensible thermal storage is low, sensible thermal devices typically have certain limitations due to their large sizes. Lucentini (2014) presented thermal storage of sensible heat using concrete modules in solar power plants. Tiskatine *et al.* (2017) carried out a detailed study on suitability and characteristics of rocks for sensible heat storage in CSP plants. For high-temperature sensible energy storage for CSP systems, the selection and analysis are done by Khare *et al.* (2013).

Table 1.4 Typical materials used in sensible heat TES storage (Mehling and Cabeza, 2008; Navarro *et al.*, 2012).

Material	Density (kg/m³)	Specific heat at constant pressure (J/kg K)
Clay	1,458	879
Brick	1,800	837
Sandstone	2,200	712
Wood	700	2,390
Concrete	2,000	880
Glass	2,710	837
Aluminium	2,710	896
Iron	7,900	452
Steel	7,840	465
Gravelly earth	2,050	1,840
Magnetite	5,177	752
Water	988	4,182

1.3.4.2 Latent Heat Storage

Latent heat storage systems including phase change-based systems have attracted more attention in fundamental research as well as in industrial applications. During latent thermal storage, the storage material is heated until it changes phase at constant temperature conditions. Latent heat energy storage allows efficient storage of thermal energy by minimizing the entropy generation in isothermal processes like evaporation or condensation. Latent heat storage systems employ the enthalpy change of a substance passing through a phase change. In CSP technology the development of absorbers directly generating steam have sparked interest in latent heat storage systems.

Latent energy storage systems have high energy density resulting in easily structured, small and flexible designs. The material that is used for latent heat storage systems is called Phase change material (PCM). Latent heat can be absorbed (charging) as well as can be released (discharging) through phase change of the PCM. The amount of latent heat stored in the material (Q) can be calculated as follows:

$$Q = m\Delta h$$

Here m is the mass of the material and Δh is the enthalpy of phase change.

Even though there are many advantages of latent heat storage system compared to sensible heat storage system, the major disadvantages of PCM are incongruent phase change, high cost, corrosiveness and less thermal stability. As a result, such materials are still in the research stage and it has not been used by industry for CSP-TES applications. There are many materials that can be used as latent heat storage materials and a list with a few important properties is presented in Table 1.5.

Raul et al. (2018) studied modelling and experimental study of latent heat TES with encapsulated Phase change materials for CSP applications. Soares et al. (2013) reviewed passive PCM latent heat TES systems. Agyenim et al. (2010) reviewed materials, heat transfer and phase change issues for latent heat TES systems. Rathod and Barnerjee (2013) carried out a comprehensive study on thermal stability of PCM used in latent heat TES systems. Further, Cárdenas and León (2013) studied design considerations and performance enhancement procedures for high-temperature latent heat TES systems.

Table 1.5 Typical materials used in latent heat TES systems (Mehling and Cabeza, 2008; Navarro et al., 2012).

Material	Melting temperature (°C)	Melting enthalpy (MJ/m^3)
Water-salt solutions	−100–0	200–300
Water	0	330
Clathrates	−50–0	200–300
Paraffins	−20–100	150–250
Salt hydrates	−20–80	200–600
Sugar alcohols	20–450	200–450
Nitrates	120–300	200–700
Hydroxides	150–400	500–700
Chlorides	350–750	550–800
Carbonates	400–800	600–1,000
Fluorides	700–900	> 1,000

1.3.4.3 Thermochemical Energy Storage

Although chemical reaction thermal storage has multiple advantages, the chemical reaction process is complex. It sometimes requires catalysers and has certain safety requirements, and there are other difficulties such as a huge one-time investment and low overall efficiency. Thus, it currently remains in the small-scale experimental stage with plenty of problems yet to be solved before any large-scale application.

The theoretical option to store energy at higher densities compared to latent heat or sensible heat storage concepts and the advantage of storing the reactants at ambient temperature make thermochemical storage systems a promising solution for longer term energy storage. On the other hand, thermochemical storage systems show a higher complexity than concepts for sensible heat storage or latent heat storage. The long-term reversibility of the reactions is an important issue. Currently, these systems are at an earlier stage of development, and there have been no commercial applications so far.

1.4 Corrosion Problem in TES-CSP System

The use of TES provides round the electricity generation. However, as seen from the literature, most of the CSP plants use TES involving molten salts at high temperatures. Such materials are highly corrosive which can damage the components of CSP. It is impossible to prevent corrosion in molten salts altogether. The effect of corrosion may be reduced by using protective coating or corrosion resistant material. To use proper and effective coating, the material must have good adhesion properties with the structural material. The coating should be uniform and less porous (Liu *et al.*, 2014). Another way to supress the effect of corrosion is by application of a cathodic potential (Schwandt and Fray, 2014). Failure due to corrosion is a major issue and such risk of failure of TES system must always be taken into account. Such cost is a function of the costs of HTF and of materials used (Liu *et al.*, 2016).

1.5 Conclusion

Integration of TES in CSP plants is important for improving environmental issues as well as alleviating the global energy crisis. This chapter provides a comprehensive discussion on the integration of Thermal Energy Storage systems (TES) in Concentrating Solar Power plants (CSP) plants worldwide. Different TES technologies and their implementation in commercial-level CSP plants are discussed. It was observed from the discussion that CSP technology is already developed for commercialization. A good number of well-established CSP plants are producing electricity worldwide, led by Spain and the United States. In addition, a good number of commercial CSP plants are under construction or in the project state hinting towards a development of the renewable energy sector in the future.

The integration of TES into CSP plants is one of the major design changes that results in significant benefits including uninterrupted electricity production even during a day without the sun radiation or nighttime. Such systems will make the CSP plant more sustainable as well as more economically competitive and dispatchable. Even though a good number of CSP plants have been using TES, fast progress is needed in the TES technologies. As observed from the current state of the art, a large number of CSP plants implement sensible thermal energy storage systems. This is due to the reliability, low cost and large data owing to sensible energy storage. However, based on the disadvantages like low energy storage density, new ways like latent and thermochemical energy storage systems may be considered for future CSP plants. Latent and thermochemical energy storage systems have larger energy storage density which might lead to a bright future for CSP-TES technology. However, more research efforts are needed to overcome the disadvantages of latent and thermochemical energy storage systems.

References

Adinberg, R., Simulation analysis of thermal storage for concentrating solar power. *Applied Thermal Engineering*, 31, 3588, 2011.

Affandi, R., Gan, C.K., Ghani, A., Ruddin, M., Performance comparison for parabolic dish concentrating solar power in high level DNI locations with George Town, Malaysia, *Appl Mech Mater*, 570, 2015.

Agyenim, F., Hewitt, N., Eames, P., A review of materials, heat transfer and phase change problem formulation for latent heat thermal energy storage systems (LHTESS), *Renew. Sustain. Energy Rev*, 14, 615, 2010.

Ahmed, R., Sreeram, V., Mishra, Y., Arif, M.D., A review and evaluation of the state-of-the-art in PV solar power forecasting: Techniques and optimization, *Renewable and Sustainable Energy Reviews*, 124, 109792, 2020.

Avila-Marin, A.L., Volumetric receivers in solar thermal power plants with central receiver system technology: a review. *Sol Energy*, 85, 891, 2011.

Barlev, D., Vidu, R., Stroeve, P., Innovation in concentrated solar power. *Sol Energy Mater Sol Cells*, 95, 2703, 2011.

Bayón, R., Rojas, E., Analytical function describing the behaviour of a thermocline storage tank: a requirement for annual simulations of solar thermal power plants, *International Journal of Heat and Mass Transfer*, 68, 641, 2014.

Behar, O., Khellaf, A., Mohammedi, K., A review of studies on central receiver solar thermal power plants. *Renew Sustain Energy Rev*, 23, 12, 2013.

Biencinto, M., Bayón, R., Rojas, E., González, L., Simulation and assessment of operation strategies for solar thermal power plants with a thermocline storage tank. *Sol Energy*, 103, 456, 2014.

Cárdenas, B., León, N., High temperature latent heat thermal energy storage: phase change materials, design considerations and performance enhancement techniques, *Renew. Sustain. Energy Rev*, 27, 724, 2013.

Cavallaro, F.. Multi-criteria decision aid to assess concentrated solar thermal technologies. *Renewable Energy*, 34, 1678, 2009.

Desai, N.B., Bandyopadhyay, S., Integration of parabolic trough and linear Fresnel collectors for optimum design of concentrating solar thermal power plant. *Clean Technol Environ Policy*, 17, 1945, 2015.

Desai, N.B., Kedare, S.B., Bandyopadhyay, S., Optimization of design radiation for concentrating solar thermal power plants without storage. *Sol Energy*, 107, 98, 2014.

Guo, J., Huai, X., Cheng, K., The comparative analysis on thermal storage systems for solar power with direct steam generation. *Renew Energy*, 115, 217, 2018.

Herrmann, U., Kearney, D.W., Survey of thermal energy storage for parabolic trough power plants. *ASME J Sol Energy Eng*, 124, 145, 2002.

Herrmann, U., Kelly, B., Price, H., Two-tank molten salt storage for parabolic trough solar power plants. *Energy*, 29, 883, 2004.

Ho, C.K., Iverson, B.D., Review of high-temperature central receiver designs for concentrating solar power. *Renew Sustain Energy Rev*, 29, 835, 2014.

Islam, M.T., Huda, N., Abdullah, A.B., Saidur. R., A comprehensive review of state-of-the-art concentrating solar power (CSP) technologies: Current status and research trends. *Renewable and Sustainable Energy Reviews*, 91, 987, 2018.

Jain, A.A., Rabi, B.J., Darly, S.S., Application of QOCGWO-RFA for maximum power point tracking (MPPT) and power flow management of solar PV generation system, *International Journal of Hydrogen Energy*, 45, 4122, 2020.

John, E., Hale, M., Selvam, P., Concrete as a thermal energy storage medium for thermocline solar energy storage systems, *Sol Energy*, 96, 194, 2013.

Kelly, B., Kearney, D., Thermal storage commercial plant design study for a 2-tank indirect molten salt system. National Renewable Energy Laboratory, 2004.

Khare, S., Dell'Amico, M., Knight, C., McGarry, S., Selection of materials for high temperature sensible energy storage, *Sol Energy Mater. Sol. Cells*, 115, 114, 2013.

Khajepour, S., Ameri, S., Techno-economic analysis of a hybrid solar Thermal-PV power plant, *Sustainable Energy Technologies and Assessments*, 42, 2020, 100857.

Kolb, G.J., Evaluation of annual performance of 2-tank and thermocline thermal storage systems for trough plants. *Journal of Solar Energy Engineering*, 133, 031023, 2011.

Liu, M., Belusko, M., Steven, T.N.H., Bruno, F., Impact of the heat transfer fluid in a flat plate phase change thermal storage unit for concentrated solar tower plants. *Sol Energy*, 101, 220, 2014.

Liu, M., Steven, T.N.H., Bell, S., Belusko, M., Jacob, R., Will, G., Review on concentrating solar power plants and new developments in high temperature thermal energy storage technologies. *Renew Sustain Energy Rev*, 53, 1411, 2016.

Llorente, G.I., Álvarez, J.L., Blanco, D., Performance model for parabolic trough solar thermal power plants with thermal storage: comparison to operating plant data. *Solar Energy*, 85, 2443, 2011.

Lucentini, D.M., Thermal storage of sensible heat using concrete modules in solar power plants. *Sol Energy*, 103, 303, 2014.

Mehling, H., Cabeza, L.F., (2008) *Heat and Cold Storage with PCM: An Up to Date Introduction into Basics and Applications*. Heidelberg, Berlin: Springer.

Mills, D.R., Morrison, G.L., Compact linear Fresnel reflector solar thermal powerplants. *Sol Energy*, 68, 263, 2000.

Morin, G., Dersch, J., Platzer, W., Eck, M., Häberle, A., Comparison of linear Fresnel and parabolic trough collector power plants. *Sol Energy*, 86, 1, 2012.

Muhammad, H.R., Arshad, S.G.N., Complementing hydroelectric power with floating solar PV for daytime peak electricity demand, *Renewable Energy*, 162, 1227, 2020.

National Renewable Energy Laboratory (NREL) Project Listing. http://www.nrel.gov/csp/solarpaces/ (visited September 19th, 2020).

Navarro, M.E., Martinez, M., Gil, A., Fernandez, A.I., Cabeza, L.F., Olives, R. and Py X. Selection and characterization of recycled materials for sensible thermal energy storage, *Solar Energy Materials and Solar Cells*, 107, 131, 2012.

Poullikkas, A., Kourtis, G., Hadjipaschalis, I., Parametric analysis for the installation of solar dish technologies in Mediterranean regions. *Renew Sustain Energy Rev*, 14, 2772, 2010.

Powell, K.M., Edgar, T.F., Modeling and control of a solar thermal power plant with thermal energy storage. *Chemical Engineering Science*, 71, 138, 2012.

Prasad, L., Muthukumar, P., Design and optimization of lab-scale sensible heat storage prototype for solar thermal power plant application. *Sol. Energy*, 97, 217, 2013.

Raul, A., Jain, M., Gaikwad, S., Saha, S.K., Modelling and experimental study of latent heat thermal energy storage with encapsulated PCMs for solar thermal applications, *Appl. Therm. Eng*, 143, 415, 2018.

Regin, A.F., Solanki, S.C., Saini, J.S., Heat transfer characteristics of thermal energy storage system using PCM capsules: a review. *Renew Sustain Energy Rev*, 12, 2438, 2008.

Soares, N., Costa, J.J., Gaspar, A.R., Santos, P., Review of passive PCM latent heat thermal energy storage systems towards buildings' energy efficiency, *Energy Build*, 59, 82, 2013.

Shouman, E.R., Khattab, N.M., Future economic of concentrating solar power (CSP) for electricity generation in Egypt. *Renew Sustain Energy Rev*, 41, 1119, 2015.

Schwandt, C., Fray, D.J., Use of molten salt fluxes and cathodic protection for preventing the oxidation of titanium at elevated temperatures. *Metall Mater Trans B Process Metall Mater Process Sci*, 45, 2145, 2014.

Tamme, R., Laing, D., Steinmann, W.D., Advanced thermal energy storage technology for parabolic trough. *ASME J Sol Energy Eng*, 126, 794, 2004.

Tian, Y., Zhao, C.Y., A review of solar collectors and thermal energy storage in solar thermal applications, *Appl. Energy*, 104, 538, 2013.

Tiskatine, R., Oaddi, R., Ait El Cadi, R., Bazgaou, A., Bouirden, L., Aharoune, A., Ihlal, A., Suitability and characteristics of rocks for sensible heat storage in CSP plants, *Sol. Energy Mater. Sol. Cells*, 169, 245, 2017.

Van Lew, J., Li, P.W., Chan, C.L., Karaki, W., Stephens, J., Analysis of heat storage and delivery of a thermocline tank having solid filler material. *J Sol Energy Eng*, 133, 021003, 2011.

Xie, W.T., Dai, Y.J., Wang, R.Z., Sumathy, K., Concentrated solar energy applications using Fresnel lenses: a review. *Renew Sustain Energy Rev*, 15, 2588, 2011.

Zapata, J.I., Full state feedback control of steam temperature in a once-through direct steam generation receiver powered by a paraboloidal dish. *J Sol Energy Eng*, 137, 021017, 2015.

Zhang, H.L., Baeyens, J., Degrève, J., Cacères, G., Concentrated solar power plants: review and design methodology. *Renew. Sustain. Energy Rev*, 22, 466, 2013.

2
Solar Thermal Power Plant with Thermal Energy Storage

Anil Kumar[1]*, Umakanta Sahoo[2] and BK Jayasimha Rathod[3]

[1]Ministry of New and Renewable Energy, CGO Complex, New Delhi, India
[2]National Institute of Solar Energy, Gwal Pahari, Haryana, India
[3]India One Solar Plant, Brahmakumaris, Shantivan, Rajasthan, India

Abstract

Energy storage systems play a pivotal role towards smooth and continuous energy supply. An energy storage system holds the generated energy for a short time and supplies it according to need. Therefore, an energy storage system is the most capable technology to meet the rising demand for energy. A device that accumulates energy is sometimes termed as an accumulator. This chapter presents a brief overview of various energy storage systems. Two experimental set ups with objective to proficient exploitation solar energy and store through solid storage systems to provide the power 24/7. A 1 MWe (3.5 MW thermal) solar power plant with 16 hours thermal storage capacity and A 1 kWe high energy density thermal energy storage for concentrated solar plant were experimented and found satisfactory results as per Indian climatic conditions. The plant operates on Rankin cycle principle. The Parabolic Reflector concentrates the solar radiation towards the in-house developed, highly efficient cavity receiver. The cavity of the Receiver which is made of monolithic cast iron acts as perfect black body and thus provides excellent thermal storage.

Keywords: Solar energy, thermal storage, power generation, solar thermal technologies

---

Corresponding author: anil.mnre@gmail.com

2.1 Introduction

In the current state of affairs, energy requirement is rising exponentially in every sector, i.e., manufacturing, infrastructure, etc. The infrastructure sector (i.e., hospitals, restaurants, lodges, shopping complexes, large educational schools, colleges, corporate offices, multiuse, etc.) is developing very fast due to growing population, need of high comfort level due to advancement in people's living standards, and rapidly increasing energy consumption. Heating/cooling requirement of the building only consumes 30-34% of the total global energy consumption. There are resulting key problems in a variety of areas like pollution control, change in climatic conditions, global warming, ozone layer depletion, etc., that create many health issues. Right now, energy management and security are the global priority topics. The Intergovernmental Panel on Climate Change (IPCC) has accepted that Greenhouse Gases (GHGs) are the first and foremost factor responsible for various environmental issues like climate change and global warming.

Large numbers of researchers suggest that promoting more use of renewable energy would be the revolutionary way to control GHG emissions. There are various renewable options available; the choice must be made according to satisfying various criteria, i.e., techno-economic, environmental issues, geographical conditions, required energy quality, etc.

Energy intensity and its 24/7 availability have become the main relative measures of countries. Energy use to Gross Domestic Product (GDP) is known as energy intensity. Its value is usually higher for developing countries compared to the already developed countries. Higher value demonstrates huge energy dependence. India consumes approximately 6% of the world's primary energy.

It can be classified as primary and secondary energy sources. Primary energy sources are normally categorized as renewable and non-renewable on the basis of their depleting characteristics, as shown in Figure 2.1. Renewable energy is derived from natural resources and they are automatically replenished. It is also known as clean energy sources.

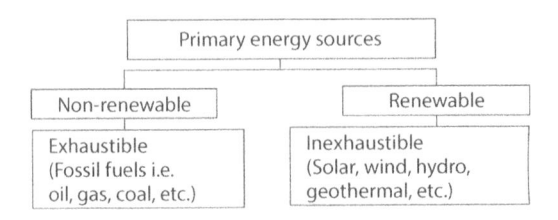

Figure 2.1 Classification of primary energy sources.

Secondary energy sources are derived from transformation of primary energy sources, i.e., heat and electricity, as shown in Figure 2.2.

Renewable energy sources are inexhaustible but as per Indian climatic conditions solar energy is the most suitable energy source, as shown in Figure 2.3.

Solar Photovoltaic (PV) and solar thermal are both efficient, getting popularity all over the globe. Selection of them depends on utility, suitability. For energy storage and 24/7 energy supply, solar thermal technology is getting more popularity than solar PV technology. Solar thermal technology is also acknowledged as Concentrating Solar Power (CSP).

- CSP Systems
  CSP systems could be cost-effectively feasible at minimum 1600-1800 kWh/m^2/year Direct Normal Irradiance (DNI) by utilizing novel technologies, substances, economies of scale and supporting renewable policies, etc. [1, 2]. CSP systems have a range of prime objectives. These are to be environmentally safe, to diminish primary cost and ground area, to increase long-term system trustworthiness, to make possible more easiness in service and maintenance. Sequestration of one ton of Carbon Dioxide-Equivalent (CO_2-e) is equivalent to one CER unit. CSP technologies can be categorized as line and point focus, shown in Figure 2.4 [3].

  Parabolic trough comprises a parabolic linear reflector. It reflects or focuses sun rays/radiations towards the receiver,

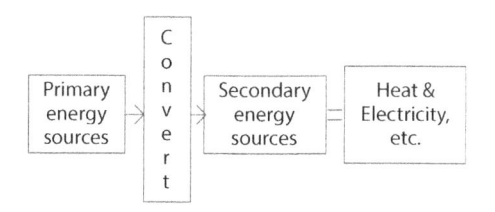

Figure 2.2 Secondary energy sources.

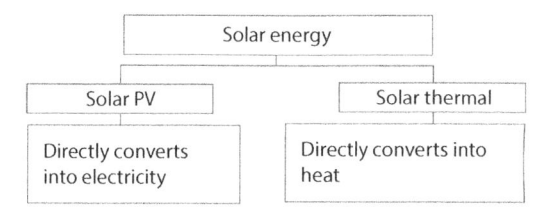

Figure 2.3 Classification of solar energy.

34 ENERGY STORAGE

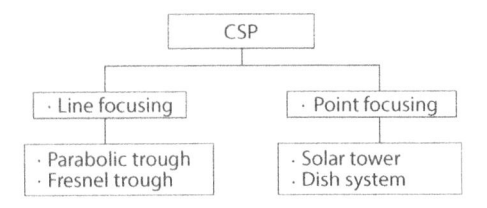

```
                    ┌─────────┐
                    │   CSP   │
                    └────┬────┘
           ┌─────────────┴─────────────┐
    ┌──────────────┐            ┌───────────────┐
    │ · Line focusing │         │ · Point focusing │
    └──────┬───────┘            └───────┬───────┘
    ┌──────────────────┐         ┌──────────────────┐
    │ · Parabolic trough │       │ · Solar tower    │
    │ · Fresnel trough  │        │ · Dish system    │
    └──────────────────┘         └──────────────────┘
```

Figure 2.4 Classification of CSP systems.

shown in Figure 2.5. Thermal heat is absorbed by working fluid (i.e., molten salt, etc.), that is filled in the receiver's tube. Reflector has a tracking system to track the maximum sun radiations. Working fluid achieves the temperature 150-350° C. Weight of parabolic linear reflector is higher due to joint less design, therefore tracking systems consume enormous auxiliary power [4].

To counter this drawback, several curved mirrors are positioned rather than a single parabolic reflector that is known as Linear Fresnel Reflector (LFR), shown as Figure 2.6. LFR is less proficient compared to parabolic trough, due to a problem in tracking its multiple curved mirrors. As a result of fixed position of curved mirrors, at the time of sunrise and afternoon cosine losses also arise in addition to longitudinal cosine losses as compared to parabolic trough.

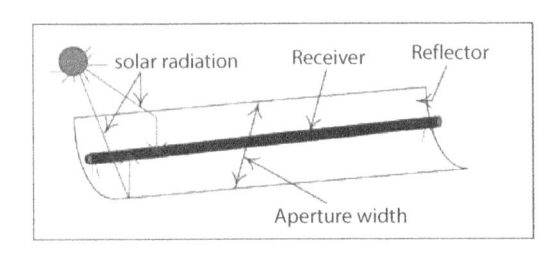

Figure 2.5 Parabolic trough collector.

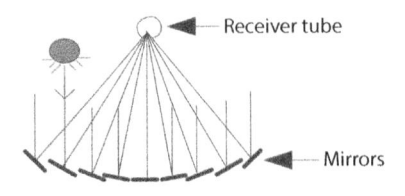

Figure 2.6 Linear fresnel reflector.

Still having these drawbacks, it is trouble-free, less maintenance and generation cost per kWh is lower than the parabolic trough [5–7]. Receiver's position is kept at a higher position to reduce its shadow effect. In contrast, due to higher height and longer path, losses also increase.

Solar tower is another technology, as shown in Figure 2.7 [8]. Array of heliostats tracks sun in double-axis. The working fluid gets heated in between 500-1000° C. The main constraint of this technology is the size, design of the tower. Parabolic Dish Collector (PDC) system concentrates sun radiations towards receiver similar to parabolic trough but focused on a single point, shown as Figure 2.8. Working fluid gets heated up to 750° C. It is more efficient compared to others but suitable for small capacity applications only.

- Energy Storage Systems (ESS)
  Rapidly rising demand of energy, fast depleting and limited stock of fossil fuels, and the serious environmental issues they create, compel a shift towards renewable energy sources. There are some critical issues while using renewable energy sources like reliability, quality, etc. Energy storage systems have the capability to solve the problems to some extent towards smooth and continuous energy supply. Due to rapid growth in the infrastructure sector (i.e. communication,

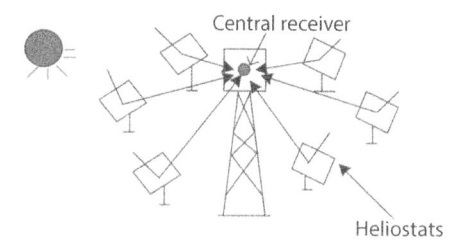

Figure 2.7 Solar tower.

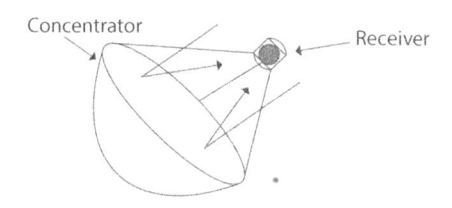

Figure 2.8 PDC system.

transport, road and rail networks, etc.) demand of energy is rising enormously, and more than 20-30% of demand is satisfied by non-conventional energy sources [9]. Renewable or non-conventional energy sources are essential for sustainable development and have many advantages over conventional energy sources like availability, environment friendly, etc. But the most important difficulty is the uneven generation of energy. Therefore, a trustworthy and affordable energy storage system becomes a prerequisite for using renewable energy [10–12]. Energy storage systems play a pivotal role towards smooth and continuous energy supply. An energy storage system holds the generated energy for a short time and supplies it according to need. Therefore, energy storage system is the most capable technology to meet the rising demand for energy. A device that accumulates energy is sometimes termed as an accumulator. There are various energy storage systems. This paper presents a brief overview of various energy storage systems. Many researches come to the conclusion that renewable energy sources are the only option for sustainable development and appropriate energy storage systems are the prerequisite. They have features to store the energy and then release it as and when required. Classification of energy storage systems are shown as Figure 2.9.

Some ESSs, i.e., flywheel energy storage, compressed air energy storage, pumped storage, batteries, regenerative fuel cell storage, and superconducting magnet energy storage are explained here. Flywheel energy storage systems store energy mechanically in the flywheel rotor by rotating the rotor. Afterward a generator is employed to convert mechanical energy to electrical. It is efficient and is used for various applications. It is preferred due to compactness, light in weight and high energy capacity. But due to limited amount of charge/discharge cycle characteristic, it is not cost-effective.

In a pumped storage system, water is pumped and stored at height during off-peak periods then utilized to generate electricity to meet the peak demand. Hydro power plants store electricity in Megawatts (MW) or Gigawatts (GW). It has many advantages, i.e., fast start-up and reliable, but it requires a large area and the cost is high. Portable batteries are well accepted in many small storage applications like transport

Energy Storage Technologies

- **Mechanical**
  1. Pumped Hydro Energy Storage (PHES)
  2. Compressed Air Energy Storage
     - Earth Air heat Exhanger
     - Stone Exhanger
  3. Flywheel Energy Storage

- **Electrochemical**
  1. Lead Acid Batteries, Advanced Lead Acid (Lead Carbon, Bipolar Lead Acid)
  2. Lithium Batteries (LCO, LMO, LFP, NMC, LTO, NCA)
  3. Flow Batteries (ZnBr, Vn Redox)
  4. Sodium Batteries (NaS, NaNiCl2)
  5. Zinc Batteries- Zn Air, ZnMnO2

- **Thermal**
  1. Sensible-
     (a) Liquid-Molten Salt, Chilled Water
     (ii) Solid-hellide, cast iron
  2. Latent-ice Storage, Phase Change materials
  3. Thermo-chemical Storage

- **Electrical**
  1. Super Capacitors
  2. Super-conducting Magnetic Energy Storage (SMES)

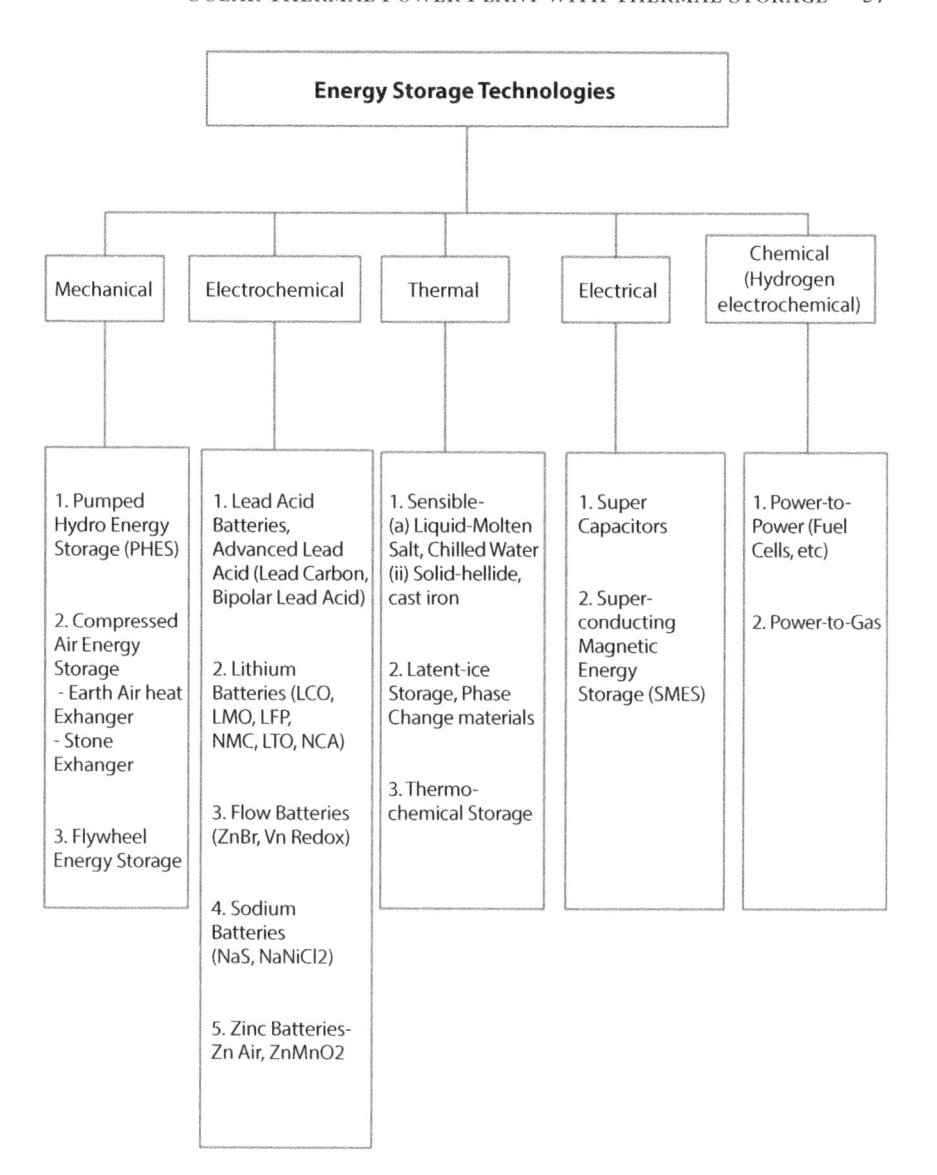

- **Chemical (Hydrogen electrochemical)**
  1. Power-to-Power (Fuel Cells, etc)
  2. Power-to-Gas

Figure 2.9 Energy storage systems. (Source: indiasmartgrid.org).

sector, utilities, etc. But it has some drawbacks like high cost, short life and a need of regular maintenance. Hydrogen fuel cell is the one type of electrochemical cell, where hydrogen is used as the primary fuel and oxygen is also required. They produce electricity with very little pollution like hydrogen cell produces by-product water. It has many advantages like no greenhouse gases, more operating time [14]. But it also has some disadvantages like facing difficulty in storing of hydrogen due to highly inflammable nature of H_2 and requirement of high capital cost due to platinum catalyst.

Superconducting magnet energy storage is an advanced energy storage system. It stores energy in the magnetic field within magnets that is developed by flow of direct current in a superconducting coil, and then releases it within fraction of cycle. Molten salt storage systems are the established commercially available concept for solar thermal power plants. Due to their low vapor pressure and comparatively high thermal stability, molten salts are preferred as the heat transfer fluid and storage medium. However, due to pricing pressure, the development of alternative, more cost-effective concepts is an important step in making thermal energy storage more competitive for industrial processes and solar thermal applications [17, 18]. A closer look at the capital cost distribution of two-tank storage systems reveals that indirect systems with a maximum operating temperature of 400 °C have differing heat transfer fluids (HTF) and storage media. For those systems, the molten salt storage media (about 35% of the direct capital costs) and the storage tanks (about 24% of the direct capital costs) are the main bearers of cost. For direct systems with operating temperatures up to 560 °C, using molten salt as the HTF and the storage media, the capital cost ratios are 34% for the storage media and 31% for the storage tank, respectively [19]. Stone storage medium is pebbles that has significantly higher thermal conductivity than normal concrete. Although the recipe of this material is quite complex the main component is quartzite, a natural geo-material readily available in many parts of the world. Further, heat is transported in and out of the storage by way of a heat transfer fluid (HTF) which flows through steel pipe heat exchangers that are cast into concrete storage elements. These elements are specially designed to deal with thermal deformations and stressing [20, 21]. Stone storage may be a good technology for CL-CSP system.

It can be concluded from comparative study of various energy storage systems that for the need of large-scale energy storage underground thermal, pumped hydro and compressed air energy storage systems are suitable. Superconductors are able to store energy with negligible losses. Fuel cells are a viable alternative to petrol engines due to their high efficiency. Flywheels have a narrow range and are suitable for small-scale operations. Molten salt and stone storage systems are gaining more acceptability for solar thermal power plants.

2.2   Literature Review

Literature on the basic concept of solar thermal systems and energy storage systems, their classifications, performances, and merits and demerits, has been reviewed. The literature review is presented under the following heads.

2.2.1   Power Installed Capacity of India

Worldwide energy demand is growing exponentially. Simultaneously environmental issues and the need to reduce the balance of fossil fuels raise the alarm and force us to switch over towards alternative options with 24/7 energy storage. Power installed capacity of India is presented in Table 2.1.

Table 2.1 Total power installed capacity of India (As on 30.09.2019) (Source: https://powermin.nic.in).

Fuel	MW	% Share
Total Thermal	2,27,644	63.2
Coal	1,96,895	54.2
Lignite	6,260	1.7
Gas	24,937	6.9
Diesel	510	0.1
Hydro (Renewable)	45,399	12.6
Nuclear	6,780	1.9
RES (MNRE)	82,589	22.7
Total	3,63,370	100

2.2.2 Energy Storage Systems

There are various options for 24/7 energy storage (i.e., batteries, solar thermal, etc.) coupled with alternative energy sources. Tremendous literatures are available for battery storage systems. Cost analysis of 1 MW solar power (i.e., Solar PV) 24x7 energy storage with lead acid batteries is presented in Table 2.2.

For MW scale solar thermal power plant based on parabolic trough collector (PTC) and molten salt as thermal storage, the following cost analysis of 1 MW solar power (i.e., Solar thermal) 24x7 energy storage is presented in Table 2.3.

It should be noted that life cycle assessment of solar PV–based power is required replacement of batteries after 5 years in case of Lead acid battery and 10 years in case of Li-ion batteries. There is no replacement of solar thermal power plant which continues to run as it maintain. Hence PV-based plant cost will be around Rs. 114 crores and solar thermal-based plant is Rs. 60 Crores for stabilized uninterrupted quality power. It can be concluded from the above analysis that considering the technical and economic benefits of solar thermal power coupled with thermal storage technology in comparison to solar PV system coupled lead acid batteries/Li-ion batteries storage technology, it is better to opt for thermal storage technology even though it is costlier during the initial capital cost. For the 20 years, considering all costs (i.e., capital plus maintenance) values of Net Present Value (NPV) and Internal Rate of Return (IRR) would be better for solar thermal energy storage systems.

2.2.3 Thermal Storage Systems

Powell *et al.* [22] observed that CSP or solar thermal power technology is suitable to be coupled with any alternative strategies like coal, natural gas, biofuels, geothermal, photovoltaic (PV), and wind. Hybridization provides high reliability, efficiency, and reduced capital costs due to resource sharing. Overall system efficiency is improved via synergy of the various resources or energy sources. An additional benefit of CSP technology is appropriate for coupling with energy storage systems, i.e., thermal energy storage (TES). Globally, four main CSP technologies are popular for power generation:

1. Parabolic Trough (PT)
2. Solar Tower (ST)
3. Linear Fresnel Reflector (LFR)
4. Parabolic Dish (PD)

Table 2.2 Costing of 1 MW Solar PV (24x7).

Load required:	24 MWh-AC
Inverter efficiency @ avg 95%	24/0.95 = 25.26 MWh
Battery @80%	25.26/0.80 = 31.57 MWh
Charge Controller @ 96%	31.57/0.96 = 32.89 MWh
Array thermal loss @ 80%	32.89/0.80 = 41.11 MWh
Radiation available	5 KWh for 5.5 hrs
PV array Capacity required	41.11 MWh/5.5 h = 7.47 MWp
Cost of PV @ 4 crore per MW	7.47 × 4 = 29.89 crores
Battery cost for uninterrupted power supply and best quality power output	30 MWh approx
Cost of lead acid battery @ Rs. 7000 per KWh for period of 5 years	30,000 × 7000 = 21 crores for period of 5 years. Therefore for period of 20 years, we have to replace the battery 5 times, so total cost for battery will be around 21 × 4 = Rs. 84 crores.
Cost of Li-ion battery @ Rs. 14000 per KWh for period of 10 years	30,000 × 14000 = 42 crores for period of 10 years. Therefore for period of 20 years, we have to replace the battery 5 times, so total cost for battery will be around 42 × 2 = Rs. 84 crores.
Hence total cost of 1 MW PV based lead acid battery operated power plant: 29.89 + 84 = 113.89 crores, say, Rs. 114 crores for stabilized uninterrupted quality power.	
So 1 MWh produces 1000 unit every hour, for 24 hrs	24 × 1000 = 24,000 unit daily
Cost of unit sale @ Rs. 5/unit	24,000 × 5 = 1,20,000
Cost of diesel plant is around Rs. 18 per unit	24,000 × 18 = 4,32,000

Table 2.3 Cost analysis of 1 MWth solar power 24x7 energy storage.

Ero trough + schott vacuum tube cost	Rs ~ 30,000 per sq.m
1 MW × 24 18% turbine	24 MWh electrical
Thermal anticipating loss 10%	110/18 = 6.1 times
Megawatt thermal capacity	24 × 6.1 = 146 MWH or say 150 MWH
Average DNI	4.5 KWh/m2
Efficiency of trough 60%	4.5 × 0.6 = 2.7 Kwh/m2 of PTC
Requirements	150 MWH = 150 × 1000 = 150000 Kwh/2.7 = 55,555 m2 of PTC × 30,000 = 16.666 crores
BOP turbine Island Storage-molten salt KNO3 + NaNO3 ~ Rs 120/Kg	
	150 MWH × 0.7 ~ 105 MWH thermal storage = 105 × 1000 × 50 ×70 = 36.5 crores storage cost
Total cost	16 cr + 36 cr = 60 crores
Hence total cost of 1 MW solar thermal with molten salt thermal storage power plant is Rs. 60 crores for stabilized uninterrupted quality power.	
So 1 MWh produces 1000 unit every hour, for 24 hrs	24 × 1000 = 24,000 unit daily
Cost of unit sale @ Rs. 5/ unit	24,000 × 5 = 1,20,000
Cost of diesel plant is around Rs. 18 per unit	24,000 × 18 = 4,32,000

Solar thermal power plant option in India is on the basis of actual Direct Normal Irradiance (DNI) of solar radiation resource assessment (SRRA) and solar Atlas of SRRA. It will also cover the possible option of thermal storage for solar power plant and 24/7 operational conditions in various parts of the country. The government of India launched the National solar mission which targeted 100,000 MW of grid-connected solar power by

2022. The majority contribution is with solar PV technologies because of widespread development and reach to grid parity as per last bidding of Rs 3 per unit cost. However, solar power plant with 24/7 operation, the same PV power plant combined with MWh batteries cost around 24-25 crores, which is quite expensive and inefficient.

In contrast, the same power with solar thermal plant with thermal storage, project cost will be lowered or at par with solar PV with stable grid reliability.

India has an opportunity to become a major contributor to development of solar thermal power. According to India Meteorological Department (IMD), clear sunny weather is experienced for 250-300 days a year by most parts of India, although it varies from region to region. India has the highest DNI range to the lowest DNI range (Ladak to Cherrapunji). Therefore, depending on the technology option available, the different regions need to sort out and propose a solution with a technology and storage option.

The different climate zone of various part of India, and suitable technology along with specific thermal storage systems is given in Table 2.4.

Worldwide, accepted thermal storage options are molten salt, stone storage, cast Iron & wrought iron storage and PCM. Lappalaineb *et al.* [23] experimented on hybrid CSP with two thermal energy storage (TES) systems at Almeria in Spain and finally validated through simulation, found good agreement between them. Two thermal energy storage systems were (i) pumping molten salt (MS) from cold tank to hot tank, and (ii) free drainage from the upper (hot) tank to the lower tank. It was concluded that molten salt was suitable for CSP and other application, i.e., nuclear power.

Bauer *et al.* [24] give an overview of commercial molten salt TES systems for CSP plants. They explained the outcome on prime decay reaction with nitrite formation in the melt and oxygen release and then a secondary decay reaction with alkali metal oxide formation in the melt and nitrogen/nitrogen oxide release. Results point out that the kinetic time constants of these two decays are not similar under the observed experimental conditions. Therefore, further future work on the nitrate salt chemistry near the stability limit is required. Cristina *et al.* [25] designed, built and tested a molten salts pilot plant at representative scale of 8 MWh_{th}. Main components, i.e., storage tanks and heat exchanger, were tested deeply. It was found that MS TES gave satisfactory results for large commercial CSP projects. Cristina *et al.* [26] observed that in comparison to sensible and latent heat storage, thermochemical storage (TCS) systems still needed to be researched thoroughly.

Table 2.4 Different climate zone of various part of India, suitable technologies & thermal storage systems.

Temperature and DNI	Parts/regions of India	Suitable solar thermal technologies coupled with storage systems
High temperature and high DNI Zone	Rajasthan, Gujarat, some part of Madhya Pradesh, Adhra Pradesh, Karnataka, TamilNadu, Bihar, Chhattisgarh, Orissa and Maharashtra.	PTC, ST, LFR and PD with all thermal storage
Moderate Temperature and low DNI	Delhi Haryana, Uttar Pradesh, some part of Bihar Chhattisgarh, Orissa and Maharashtra, West Bengal and North East Region.	PTC, ST, LFR and PD with molten salt thermal storage
Low temperature and high DNI Range	Leh, Ladak, Kargil and some part of J&K.	Parabolic Dish (PD) with cast iron storage

2.3 Energy Demand of World

Firdaus *et al.* [27] briefly discussed in their paper that the energy demand of the world represented huge potential for solar energy. However, it plays only a limited role in present conditions. It consists of the up-to-date generation of the concentrator and on how the other parameters compare and considers the affordable one. Each and every generation of concentrator are compared simultaneously with each other and further optimization in concentrator had also been an understanding for future scope.

- Concentrator
  A. Dang [28] discussed the existing concentrating solar technology. He also analyzed the particular concentrator designing, manufacturing, modification, testing, tracking mechanism, and material in use, optical and thermal performance. Some modification had also been provided to enhance the performance of a concentrator. The different

focusing concentration technology and direction vary calculation had been shown.

T. Cooper *et al.* [29] narrated about the line to point focus concentrator by maintaining the tracking mechanism on a single axis and reducing the cost of a collector with the help of such design consideration for the primary concentrator; this can be achieved while allowing the thermodynamic concentration limit. The proposed solar concentrator design is well suitable for large-scale application with the concentration ratio 500-2000.

D. Gadhia *et al.* [30] worked on the development of parabolic solar collector with such inputs and collaboration. The slight growth in solar technology the food processing technology and all other applications. The approved technologies and its commercialization were successfully installed and benefits were realized. Apart from that, the feedback of this technology was also given by their users. The clean development management and formation of eco-friendly environment will safeguard us and also our planet earth.

- Receiver:

T. Lee *et al.* [31] proposed their motive about the design optimization of a solar tubular receiver to rise up to working performance. The poor condition which was affecting its performance was analyzed. The multiple factors affecting the system performance were the length of the receiver, the maximum number of inlet pipes, porousness and the thermal conductivity of that porous medium. The maximum possible effect in each variable on certain physical conditions like maximum temperature and pressure drop was identified and the suitable optimal design was sorted out. Thus the rare design was suggested to grow up the factor of efficiency. Further, this alternate design will overcome its performance through the manufacturing process.

A.L. Avila-Marin [32] completely reviewed the volumetric receiver design to optimise minimal heat loss. The author gave a comparison between the volumetric receiver and tube receiver with their different working principle and geometry. The volumetric receiver includes the porous material which easily absorbs the highly concentrated radiation inside the volumetric structure. Then further the heat is transferred to the fluid passing through the same structure. It had been widely used in central receiver system technology. Thus the study concludes with discussion of the pre-existing and up till date volumetric receiver in use; analysis was made of various parameters and the best configuration was considered.

R. Duggal *et al.* [33] numerically identified the three-dimensional models of trapezoidal cavity receiver which is in use in linear Fresnel reflector with water as a heat transfer fluid was analyzed by both. They also fully described the 3D model. The issue of thermal loss was predicted and certain parameters were noted down and the effect was also seen with variant losses. On the other hand, the suitable design considerations were analyzed and an alternate receiver was proposed for further use.

- Thermic Fluid
B. Gobereit *et al.* [34] offered a complete summary of the computational fluid dynamics model of a particle receiver and gave profound information regarding various known factors related to it, and also it is compared with the pre-existing prototype. The thermal radiation losses predicted by the CFD model are totally different as per the estimated one. It is a true concept for increasing the system efficiency for various solar thermal applications. M.J. Bustamante [35] described the heat transfer fluids transferring and utilizing the collected solar heat with the help of solar thermal energy collectors. The solar thermal collectors are categorized according to the temperature range, namely low, medium and high. Low-temperature solar collectors use phase changing refrigerants and water as heat transfer fluids. The uses of water-glycol mixtures as well as water-based nano fluids are obtaining momentum in low-temperature solar collector applications. The hydrocarbons are also used as refrigerants in many cases. In medium temperature solar collectors the heat transfer fluids include water, water-glycol mixtures, i.e., trimethylene glycol (green glycol) and also naturally occurring hydrocarbon oils in various compositions such as aromatic oils, naphthenic oils, and paraffinic oils in their increasing order at suitable operating temperatures. In high-temperature solar collector, the synthetic hydrocarbon oils as heat transfer fluid are used as a fluid of choice in wide applications while other heat transfer fluids are being used with varying degree of experimental maturity

and commercial viability – for maximizing their benefits and minimizing their disadvantages.

A. Sinha [36] reviewed the concentrating solar technologies with the heat transfer fluid which are recently in use in India. The various kind of heat transfer fluid possess variant physical and thermal properties. As per our needs, the concentrating solar power plant had formed an alternative source in power generation around rural field area. The installed concentrating solar power plant includes the heat transfer fluid Therminol VP-1, Synthetic oil, Dowtherm A, etc., in use. Heat transfer fluid like Hitec salt, dowtherm A are very stable compounds and are used for power generation limited to 700 °C. Some alternate sources of energy rather than coal are essential to meet the gap between demand and supply. Through this paper, the author discussed the various properties of each heat transfer fluid with reference to concentrating solar power plant in India. He also suggested the halide-based salt utilization in these plants for small-scale power generations in rural areas. To utilize the phase change material mainly for two different motives, i.e., primary for working of the heat transfer fluid and secondary regarding thermal energy storage.

- Solar Thermal Energy Storage

A. Sharma *et al.* [37] gave a successful review regarding the thermal energy storage system with the phase change material concept, and also its applications had been discussed in this paper. They discussed the latent heat storage system with PCM having a dominant way of storing. There is an advantage of high storage density and isothermal characteristics in this particular storage. There had been a number of applications with the PCM with latent heat storage. The various phase change material had been studied and analyzed about its melt fraction and thermal properties. This paper also summarizes the investigation and analysis of the available thermal energy storage systems incorporating PCMs for use in different applications.

B.P. Jelle *et al.* [38] discussed in their paper on thermal energy storage system according to phase change material which can lower the energy consumption of the buildings. The releasing and storing of heat in a certain temperature limit, the inertia of the building increases and the room temperature becomes stable. The maximum amount of energy is stored at a high temperature and returns back at a certain temperature due to an increase in thermal mass at a narrow temperature range. A high potential of energy had been saved but not yet that much optimized for building application purpose. Some known materials with a transition around comfort temperature, and those existing do have a relatively low heat of fusion. S. Kuravi *et al.* [39] discussed the thermal energy storage technologies and the factors which are to be considered at different levels in concentrating solar plants. The thermal energy storage is the vital component in concentrating solar plant for increasing the efficiency and the performance. There is a lack of storage system in solar power plant and the designing along with the integration of the storage system is not so highly focused to build it. The study of the various thermal energy storage system is pointedly specified and other economic aspects are also summarized. Thus the arrangement of the various storage system is required to achieve the expected efficiency. S. Khare *et al.* [40] discussed the alternative source apart from the conventional energy. The suitable non-conventional energy is solar energy which is eco-friendly in use. The solar power in India is developing day by day. It is easily available everywhere; also the demand of various sectors through solar energy is increasing widely and if we want to save our environment then we must use solar as a part of the solution. In this paper, the schemes available in India, solar mission, a case study of some applications, etc., is discussed shortly.

2.4 Experimental Set Up

A solution towards energy security and various environmental issues is the adoption and promotion of renewable energy systems. Solar energy has the tremendous scope of energy generation, i.e., electricity and heat, both. In this chapter, the methodology adopted for CSP system with energy storage system is presented. The adoptability of a solar thermal system can significantly be enhanced by coupling an energy storage system. The title of the first experimental set up is "1MW electrical (3.5 MW thermal) solar power plant with 16 hours thermal storage capacity". The aim of the experimental set up is to establish a 1 MW capacity solar thermal power plant with

16 hours storage facility based on Parabolic Solar Reflectors at an estimated solar to electric efficiency of about 12%. The title of the second experimental set up is "high energy density thermal energy storage for concentrated solar plant".

2.4.1 Description of Experimental Set Ups

Experimental set up I is established near Shativan Campus, Bhrahmakumaries, Talheti, Abu Road-307510, Rajasthan.

(a) Design parameters of the experimental set up I:
Parabolic Solar Reflector (PSR) - 60 SQM PSR is presented as Figure 2.10. It is completely designed with space frame comprising solar grade curved mirrors. Mainly it has four parts: (i) Supporting stand, (ii) Rotating wheel, (iii) Central bar space frame, and (iv) Outer frame with cross bars and long bars.

All the materials used for the PSR are of mild steel grade as per IS 2062 and as per IS 4923 for solid and hollow sections, respectively. All the surface areas are well coated with paints (epoxy paint and PU paint) for the long life service.

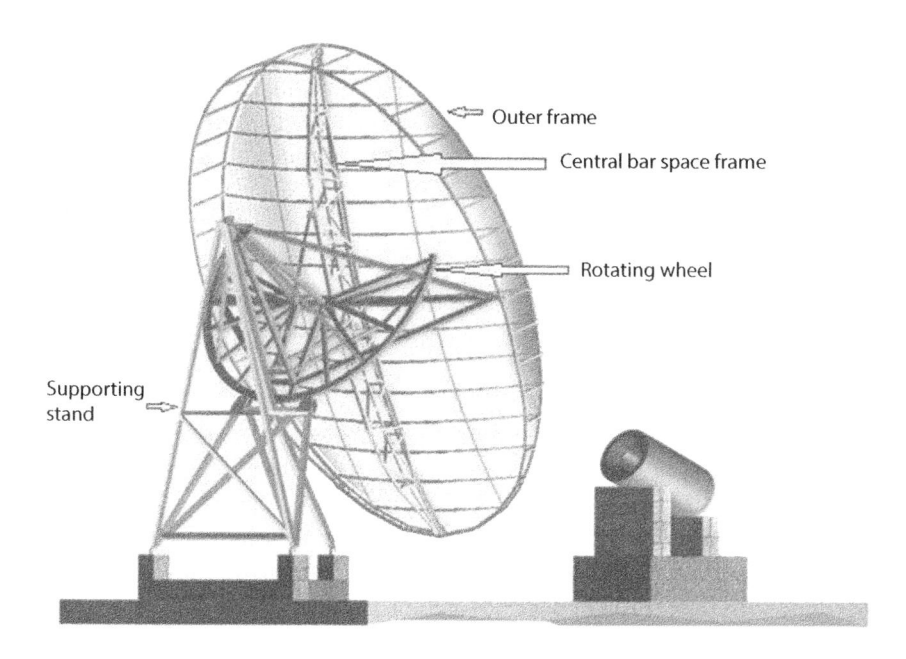

Figure 2.10 Parabolic solar reflector.

- Main parts of Parabolic Solar Reflector:
  Supporting Stand: Concrete foundation is done for providing proper strength to 60 SQM Parabolic Solar Reflector. Designing of stands is done according to latitude of the location, hence different design is done for different locations. Base of supporting stand is selected in triangular shape for providing proper strength with less land requirement.
  Rotating Wheel: It links to supporting stand and parabolic frame, as Figure 2.10. Materials employed for rotating wheel are of mild steel grade as per IS 2062 and M.S. sections as per IS 4923. The daily tracking arrangement is through rack and pinion arrangement with actuators and DC motor for daily rotation.
  Parabolic Outer Frame: The next main component of the Parabolic Reflector is the parabolic frame. The outer frame is designed in three parts as per the requirement of flexibility and rigidity balance to accommodate various shapes of parabolas for different seasonal requirements. This outer frame provides hinge support to various cross-bars and long-bars that are designed to support the mirror pieces that make a perfect parabolic reflective surface, as Figure 2.10.
  Central Bar Space Frame: It is a backbone of the outer parabolic frame. This is used for tracking mechanism on the longitudinal axis of the parabola. This has three mechanical actuators that are driven by DC motors. This structure, due to its flexible behaviors, enhances the concentration ratio of the output focus at the focal point.
  Flexible Parabola: The structural design has the flexibility option in the structure to provide flexible parabola, presented as Figure 2.11. There are different parabola options for different seasons. These flexible parabolas are possible through automatic dual axis tracking mechanism. There are three types of tracking systems:

  (i) Daily tracking: This tracking mechanism allows the Reflector to track the sun throughout the day. This is done with the help of rack and pinion mechanism.
  (ii) Seasonal tracking: This tracking mechanism allows the Reflector to align with the changes in the angle of the sun due to change in the season.

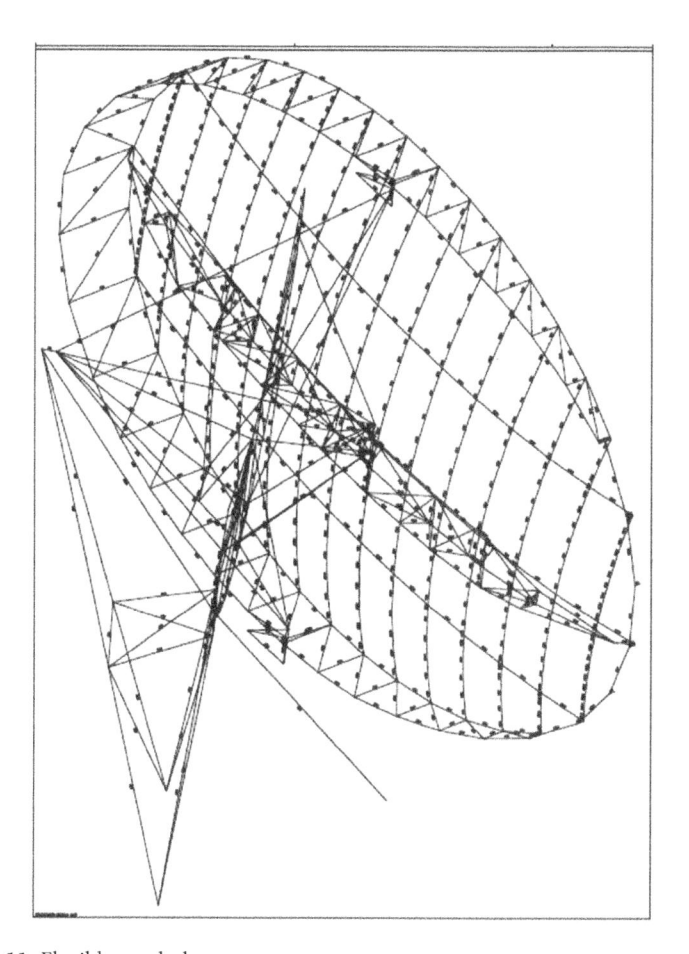

Figure 2.11 Flexible parabola.

(iii) Shape change tracking: This tracking mechanism allows the reflector to change the shape of the parabola to increase the concentration ration of the focus.

All the three types of tracking systems are fully automatic coupled with mechanical actuator (D_1) and shape change actuators (S_1, S_2, S_3, S_4), DC motors, micro controller and programming systems, that is all referred to as "IMATRACK POWER", as Figure 2.12. It tracks to 60 SQM PSR.

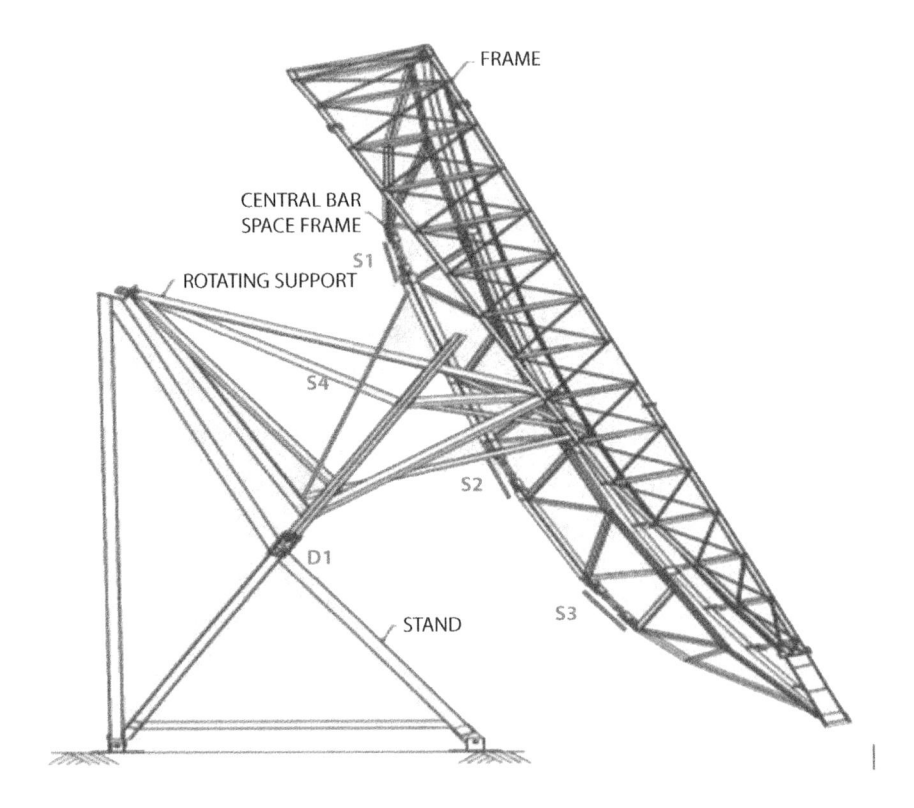

Figure 2.12 Various actuators.

- Static Cast Iron Cavity Receiver:
  Receiver is constructed by monolithic cast iron, in conical cavity shape, opening of 500 mm and 700 mm deep in conical design. A single helical boiler grade coil is wound around the monolithic conical cavity cast iron body around the periphery, as Figure 2.13.

  The whole body of receiver is shielded by mineral wool insulation of 6-inch thickness with aluminum cladding to decrease heat loss to air. Receiver is mounted on a triangular structure with fixtures to provide stability with minimum heat loss.

(b) Experiment set up II

Experimental set up II is established at Ram Krishna Dharmarth Foundation (RKDF) University, Bhopal. Details of equipment used in experimental set up II (as shown in Table 2.5).

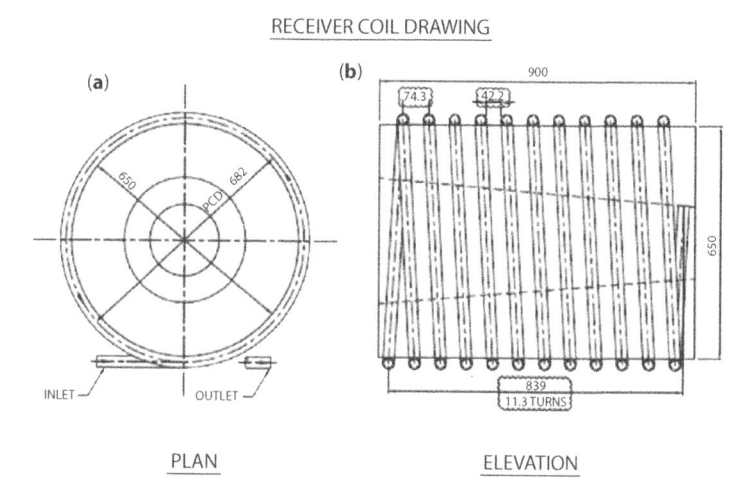

Figure 2.13 (a) Plan of receiver coil drawing (b) Elevation of receiver coil block drawing.

Other materials are;

1. Thermocouple R-Type: 360mm Long; 0-1500 °C range; Compensating Cable SS braided and Digital Meter along with accessories.
2. Thermo-Electric Device duly designed and fabricated at Micro-and nano Fabrication Clean Room (MNCR), RPI, USA. Conversion Efficiency more than 5%.
3. Thermal Storage Salt for Research developed at the RPI Lab, USA and imported vide invoice dated 25th April, 2016 – having high "Energy Density" exceeding 300kWh/m^3 and Density 2200 kg/m^3.

The aim of this project is to demonstrate a solar thermal storage system with 1 kW capacity of volumetric energy density, exceeding 300 kWh/m^3 and capable of operating at high temperatures up to 1000 °C. In comparison, the volumetric energy storage density for water is typically around 80 kWh/m^3 and 200 kWh/m^3 for molten salts used in solar thermal plants. The unique aspects of this system are the selection of an alkali halide salt with high melting temperature and a corrosion-resistant, low-cost ceramic container material. Flux grown crystals of mixed alkali halide compounds doped with metallic impurities shown as Figure 2.14. The thermal storage unit is coupled with a high solar concentrator system

Table 2.5 Equipment list used in the experimental set up.

Sl. no.	Name of equipment	Specifications	Make/Model	Cost (FE/Rs)	Date of installation	Utilization rate (%)	Remarks regarding maintenance / breakdown
1.	Solar Tracker Unit	Tracker Unit Assembly complete with Micro-Controller with Computer Interfacing Facility and full auto operation throughout the year and data logging system.		7,04,175.00	23rd Jan,2016	100%	Control System Components Replaced
2.	Heat Transfer Unit (Boiler Feed Pump & Piping & Core of MS and Copper)	Plunger Type Positive Return Metering; 0-50 LPH with discharge pressure 11kg/cm^2; Flooded suction; Pump speed – 145 RPM; All material SS 316 ; Plunger – Hard Chrome Plated	D4#23P SR NO: CE-5606 Dosing Metering Pump	1,19,914.85	26th Mar,2016	75%	Piping work was revised for better steam generation and the Core is under revision to Cast-Iron crucible design for improving steam parameters.
3.	Core Material from RPI USA	Mild Steel, Copper and Cast-Iron		3,38,401.00	7th May,2016	75%	Size of Crystal revised
4.	Fresnel Lens	1100mm(dia.) x 5mm(thick); MMA Polymer; Focal Length of 1300mm	HS 71171900000 Focal length 1300mm 1100mm Dia.	31,000.00	23rd Jan,2016	100%	-

Figure 2.14 Size and photo mixed alkali halide compounds doped with metallic impurities.

(1000 – 10,000 Xs). In this project, we propose to develop and demonstrate an affordable, high energy density thermal storage system that can store heat at temperatures around 1000 °C. Cast-iron core crucible design and schematic are presented as Figures 2.15 and 2.16, respectively. Solar thermal storage with solar tracker unit is shown as Figures 2.17 and 2.18. The tracking motor and tracking chain are shown in Figures 2.19 and 2.20.

2.5 Experimental Data Analysis, Results and Discussions

CSP coupled with storage system is a promising technology for 24/7 energy supply. Experimental results of both the arrangements are satisfactory. It is observed from experimental results that solid storage systems are the appropriate option.

ALL DIMENSIONS ARE IN MILLIMETERS (mm)

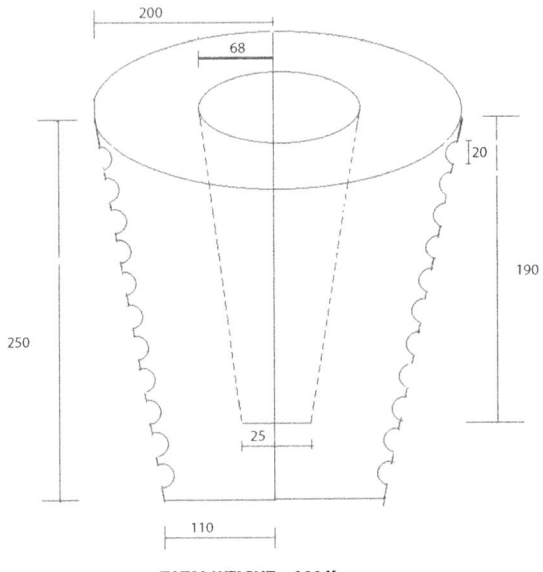

TOTAL WEIGHT = 100 Kg

Figure 2.15 Cast-iron core crucible design.

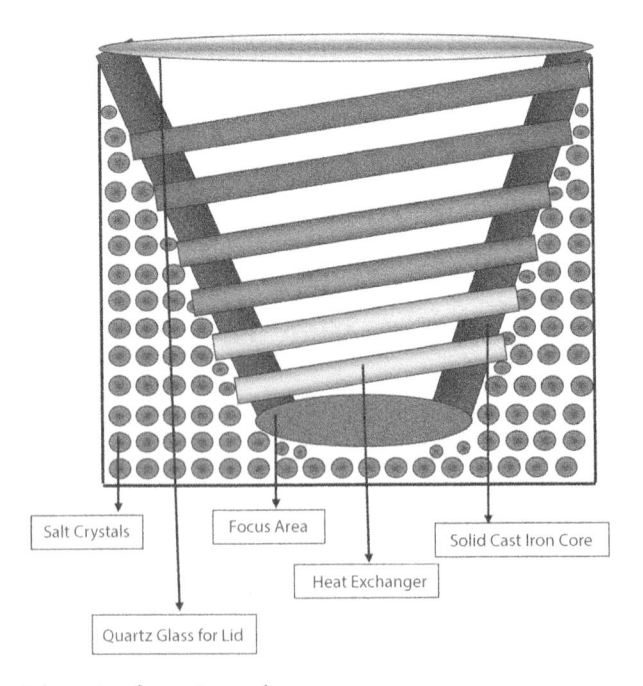

Figure 2.16 Schematic of experimental set up.

Figure 2.17 Installation of the solar thermal storage and solar tracker unit.

Figure 2.18 Solar thermal storage with solar tracker unit.

58 Energy Storage

Figure 2.19 Tracking motor.

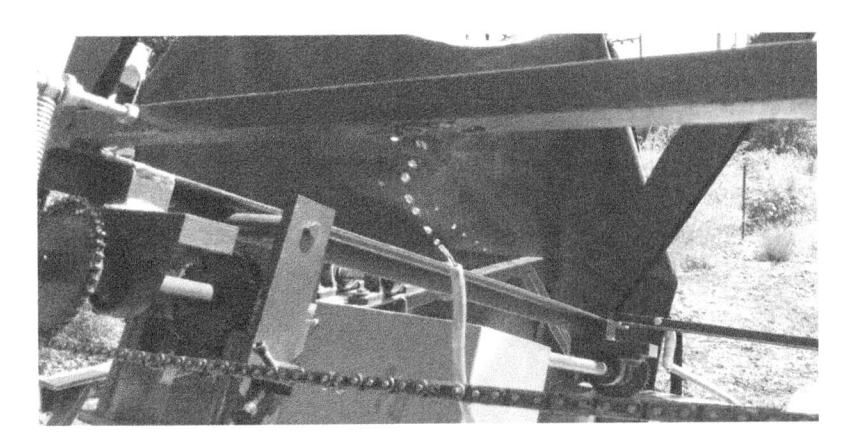

Figure 2.20 Tracking rope & chain.

2.5.1 Performance of Reflector Round the Year (Experimental Set up I)

The aperture area of the reflector is as shown in Figure 2.21; changes occur in different seasons as per the inclination with the polar axis.

- Detailed Project Report (Experimental Set up I)
a. Experimental works carried out:
   First Prototype Receiver testing set up and result:

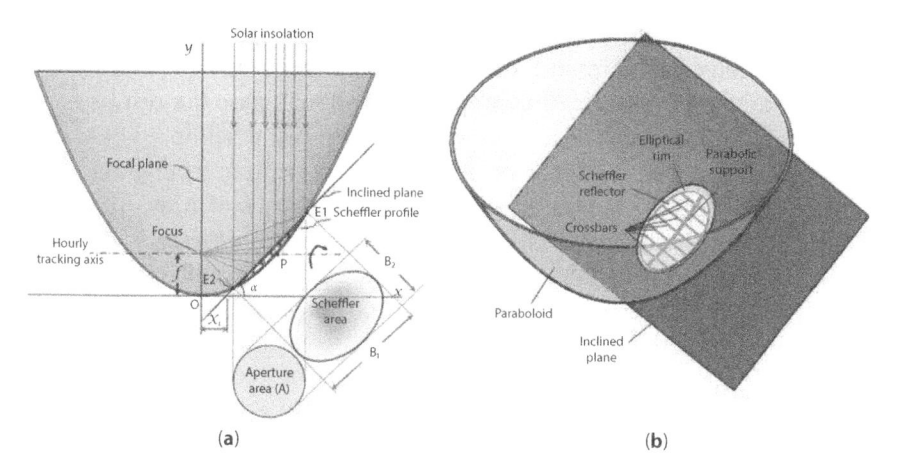

Figure 2.21 Changes in aperture area of reflector.

Thermal Storage Receiver Testing & Results:

i. Measuring system setup for first Prototype: The schematic diagram explains the set up of the measuring system to evaluate the 60m^2 Paraboloidal Reflector and the static cast iron receiver. Components involved in the Measuring system:

ii. 60 SQM Paraboloidal Reflector 4 nos. of Temperature sensors Type K embedded in the Receiver

iii. Pressure sensor and temperature sensor to measure steam parameters

iv. Campbell data logger to record the measurements (temperature)

v. Weather station (Shadow ban redo meter, Direct normal radiation DNR, Wind Speed, Wind direction, Air temperature)

The measuring systems consist of two subsystems: Subsystem one collects the data of the receiver, while subsystem two is the heart of the weather station; both subsystems collect data through Campbell CX 1000 data loggers Receiver system. The measuring system of the receivers consists of four receiver temperature sensors (type K thermocouples), a water meter, and steam pressure sensors as well as a temperature sensor for the water and for the steam (type K thermocouples). The water meter allows the manual reading of

the water flow and the amount of water injected into the receiver. The pressure sensor is from Siemens pressure transmitter of required pressure range. Following are the results obtained from the tests carried out with below set up:
- The thermal behavior of the Receiver without front glass
- The testing was carried out in the month of May. The Static receiver is charged during the daytime through the solar rays reflection focus from the parabolic reflector.
- The Receiver is charged without front glass covering.

a) Simulation report of thermal behavior of 5 no. of Receivers in series: Simulations were done in the "ColSim" simulation environment, for the cast-iron receiver in series connection to determine the thermal behavior pattern and to determine the thermal storage duration for further simulation of the plant:

The following observations and assumptions were considered in the simulation of 5 receivers in series:

- The functioning of the Receiver model considered for simulation has been matched and the parameters of the Receiver model were adapted to fit the actual results obtained from testing the first prototype receiver.
- In this simulation receivers with equal mass in series of 5 nos. are considered.
- Actual averaged Weather data for three years is considered for the timeframe used in this simulation report.
- Various other assumptions for mass flow rate of fluid (in this case water), with receiver front glass position are studied for thermal storage behavior.
- The maximum mean temperature achieved by charging the receiver is assumed as 500 °C, while the minimum temp till discharge the receiver is assumed as 260 °C.
- The receiver geometry is idealized to cylindrical structure geometry and is divided into equally spaced different subsections of the cylinder representing different temperature gradients along with the length of the heat transfer coil.
- The incoming radiation, the absorbed radiation, the heat transfer between different subsections and the heat transfer coil, the ambient losses are considered and matched in the simulation results according to the actual achieved test results of single prototype receiver.

Solar Thermal Power Plant with Thermal Storage 61

To summarize, the assumptions in this study are:

- One-dimensional heat transfer.
- Constant heat conduction coefficient for cast iron (line system).
- Every node has same amount of volume.
- Numbers of turns in helix are equal to the number of nodes.
- Reflections inside the cavity are neglected.
- There is no temperature gradient from cavity to outside diameter of the cylinder.
- Thermo physical properties of fluid are the same at every point inside the specified pipe section.
- Storage is composed of a cylindrical block of cast iron. To analyse the heat transfer between the cast iron nodes, the block is divided into 20 cylindrical shaped nodes. The results for the temperature distribution in the storage under specific operation temperatures is shown in Figure 2.22.

b) Heat Loss and Radiation for different Receivers in a module:
Figure 2.23 shows the heat losses to ambient (Q_loss, upper graph) and the heat gain through absorbed radiation (Q_rad, lower graph) of the first and last receiver in a module. The heat loss in the last receiver is significantly higher (almost by a factor 2) than the loss of the first receiver due to the lower average temperatures of the first Receiver. The first Receiver can

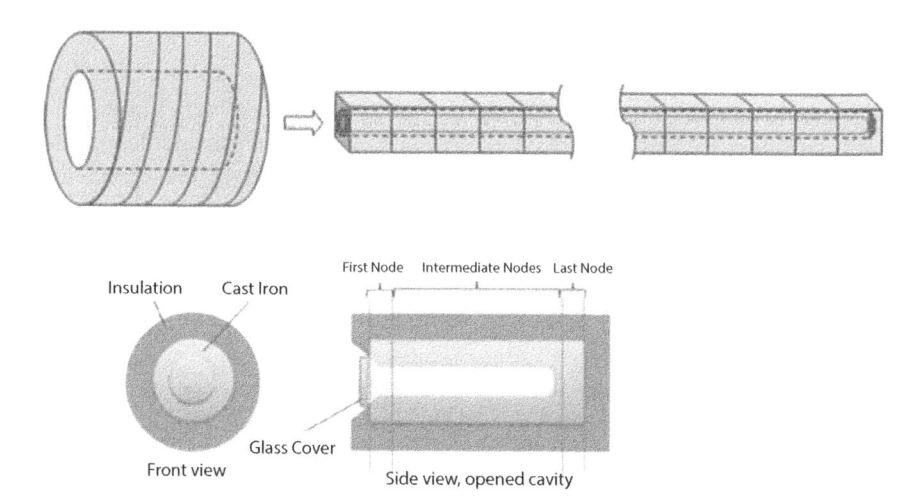

Figure 2.22 Results for the temperature distribution in the storage under specific operation temperatures.

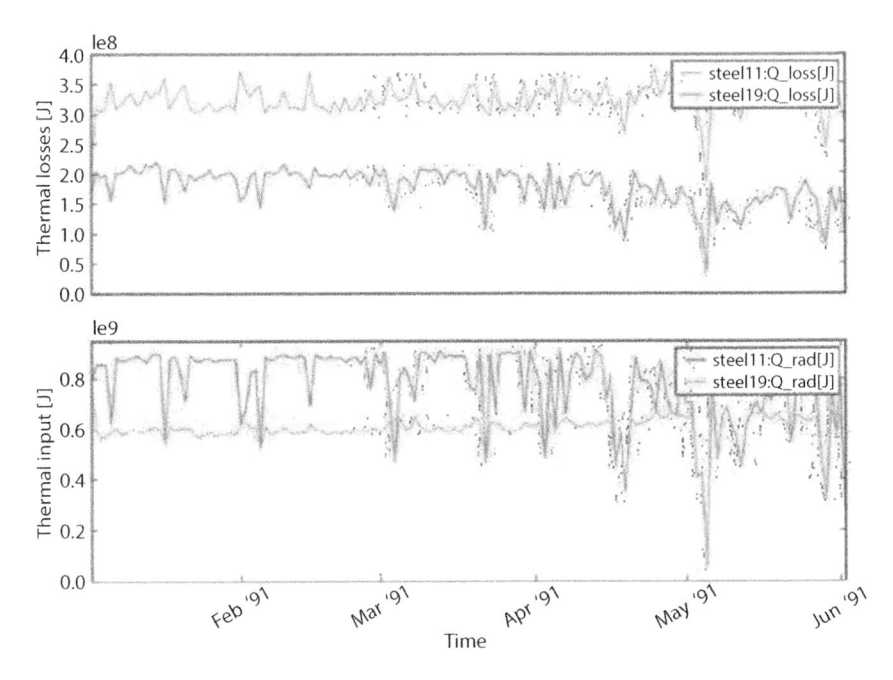

Figure 2.23 Heat loss and radiation for different receivers in a module.

accept up to 35% more incident radiation (as compared to the last receiver) until the overheating protection requires to cover the Receiver door. Le8: Thermal losses – heat losses; Le9: Thermal input – heat gained. Orange curve is for first Receiver in Series and Green Curve is for last Receiver in series.

c) Simulation report of thermal behavior of 10 no. of Receivers in series:
Simulations were done in the "ColSim" simulation environment, for the cast iron receiver in series connection to determine the thermal behavior pattern and to determine the thermal storage duration for further simulation of the plant:

The following observations and assumptions were considered in the simulation of 15 nos. of receivers in series:

- The functioning of the Receiver model considered for simulation has been matched and the parameters of the Receiver model were adapted to fit the actual results obtained from testing the first prototype receiver.
- In this simulation receivers with equal mass in series of 15 nos. is considered.

- Actual averaged Weather data for three years is considered for the timeframe used in this simulation report.
- Various other assumptions for mass flow rate of fluid (in this case water), with receiver front glass position are studied for thermal storage behavior.
- The maximum mean temperature achieved by charging the receiver is assumed as 500 °C, while the minimum temp till discharge the receiver is assumed as 260 °C.
- The receiver geometry is idealized to cylindrical structure geometry and is divided into equally spaced different subsections of the cylinder representing different temperature gradients along with the length of the heat transfer coil.
- The incoming radiation, the absorbed radiation, the heat transfer between different subsections and the heat transfer coil, the ambient losses are all considered and matched in the simulation results according to the actual achieved test results of single prototype receiver.

2.5.1.1 Simulation Results

To quickly give an indication on the effect of higher masses, reduced heat loss.

Only a short period of time of 12 days of January has been simulated. The variability of results is shown in the figure below. From this we conclude that we may choose the day Jan. 7 as a representative "good day" for this month. The detailed evaluation and comparison of cases is done only for this day and presented in the following. Below table Accepted radiation, thermal losses and thermal gains of all receivers (1 15) in the module (for Jan. 7) and their relative contribution to the total radiation/loss/gain of the module. All receiver masses are equal, 4100 kg. Equal distribution would be a contribution of 6.7% (1.15%) for each receiver. After the successful testing of the first prototype Receiver in May 2011 and conducting simulations through the simulation software "ColSim" simulation environment for both single receiver as well as 5 nos. of Receiver sin series and subsequently 15 nos. of Receivers in series as discussed above; it was clearly evident that we shall go for monolithic cast iron receiver around the conical cavity from all the three sides, with proper insulation to minimize the thermal losses. Accordingly, the testing set-up was set for practical testing and measurements to further explore different heat transfer design concepts and to confirm the results seen in the simulation results for 15 nos. of Receivers in series. The Receiver Charging-Discharging is shown

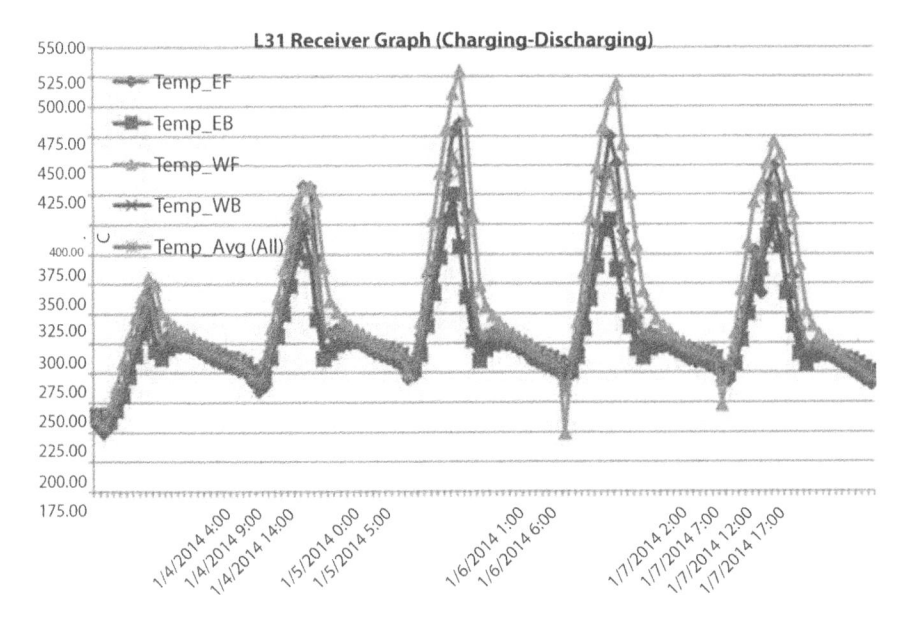

Figure 2.24 Receiver charging-discharging.

in Figure 2.24. Here below, we share the testing results for all the testing works carried out in two phases:

1. Single monolithic cast Iron cavity Receiver with different heat transfer design concepts
2. 10 nos. of monolithic cast iron receivers in series with different mass flow rates

L31 Receiver features 'Receiver with heat transfer paste':

1. Conical cavity receiver
2. Weight – 3710kg
3. Length – 1 meter
4. Coil length – 31 meters,
5. Coil Pitch – 75 mm. Coil is wrapped around the casting, heat transfer paste was applied to fill the gaps between cast Iron block and coil.
6. Temperature sensors detail:
7. Temp EF – Temp east front, Temp EB – Temp east back
8. Temp WF – Temp west front, Temp WB – Temp west back

L21 Receiver features 'Receiver with grooving':

- Conical cavity receiver
- Weight – 3610 kg
- Length – 1meter
- Coil length – 31 meters,
- Coil Pitch – 75 mm. casting was machined to have groove made of the exact size of the coil diameter, coil was fitted inside these grooves. (Photo 4, Photo 5)
- Temperature sensors detail: Temp EF–Tempeast front, Temp EB–Tempeast back Temp WF – Temp west front, Temp WB – Temp west back. The receiver with grooving is shown in Figure 2.25.

The plant operates on Rankine cycle principle. The Parabolic Reflector concentrates the solar radiation towards the in-house developed, highly efficient cavity receiver. The cavity of the Receiver which is made of monolithic cast iron acts as perfect black body and thus provides excellent thermal storage. The boiler grade coil around the body acts as a heat exchanger which allows for water to exchange heat and convert into steam. The thermal storage can be operated between 250 °C to 550 °C and can be discharged on demand. The steam generated is mostly superheated steam and the rest is saturated steam at operating pressure from 38 bar to 44 bar gauge

Figure 2.25 Receiver with grooving.

pressure. The total field is divided into 23 nos. of modules. Each module has receivers connected in series. The plant is designed as a captive plant (off grid) as per demand to provide electricity to the headquarter campus Shantivan at Abu Road, Rajasthan.

2.5.1.2 Typical PID of a Solar Module from 'India One' Solar Power Plant

A typical module consists of multiple cast iron cavity Receivers in series, the number of Receivers in series depends on the layout and land availability. The no. of Receivers ranges from minimum 13 numbers to maximum of 45 numbers in series, each module (each row) is given Alphabet for identification starting from "A" from the north of the lay out to "W" ending towards the south of the layout. Each module is aligned in exact east-west direction facing south and receiving solar radiation on the reflective surface. Since there are 23 nos. of modules in the field, therefore there are also 23 nos. of piston pumps (positive displacement pumps), i.e., one pump per module. Each module acts like a boiler that delivers superheated steam to the steam header. The first 35% of Receivers in series in a module act like economizer that heats up water into steam, the next 35% of Receivers in series in a module act to generate saturations team and the last 30% of Receivers act like super heaters to generate superheated steam which is then finally connected to the steam header. The required operating pressure of steam is delivered by piston pump at the inlet of the module. The mass flow rate of the pump is controlled by VFD (variable Frequency drives) that take their input signal of operation from the centralized pressure control mechanism which in turn takes signal throughout put frequency of generator. The plant is off-grid connected to the Brahma kumar campus that has load demand that can cater to 25,000 people for lodging and boarding. Hence, the load demand is indirectly transferred to the VFD of the pumps which synchronizes the demand with the supply and accordingly varies the mass flow rate of the water which then converts into steam and goes to the steam turbine. The mass flow rate for each module is programmed as per the length of the module, i.e., number of Receivers in series for the module. The bigger the modules the bigger is the flow rate through it. In this way, the complete solar thermal field with each module has synchronized thermal behavior of charging and discharging. Each module is installed with analog as well as digital instrumentation that reads the important parameters related to pressure, temperature. Also, each module is treated as a boiler and accordingly, each module has a dedicated safety valve. Also, the module is equipped

Solar Thermal Power Plant with Thermal Storage 67

with strainers, Isolation valves and Non-return valves wherever necessary for smooth operation and maintenance. All modules have common water header on inlet side and common steam header on the outlet side. Thus, the modules derive water from the common water header and supply steam generated to the common steam header.

2.5.1.3 Quantity of Steam to Turbine

The steam Turbine allows a certain quality of steam for its functioning it requires steam of pressure ranging from 38 Bar to 44 Bar and temperature ranging between 252 °C to 410 °C, respectively. And minimum quantity of steam starting from 4000 kgs to 6500Kgs; therefore it is very important that the solar thermal field generates steam of the required quality and quantity of steam. In order to derive this steam form the solar field, PID logic is implemented in the control logic that ensures the requirement. The PID logic implemented in the control logic is shown in Figure 2.26 and operating parameter of turbine is given in Table 2.6.

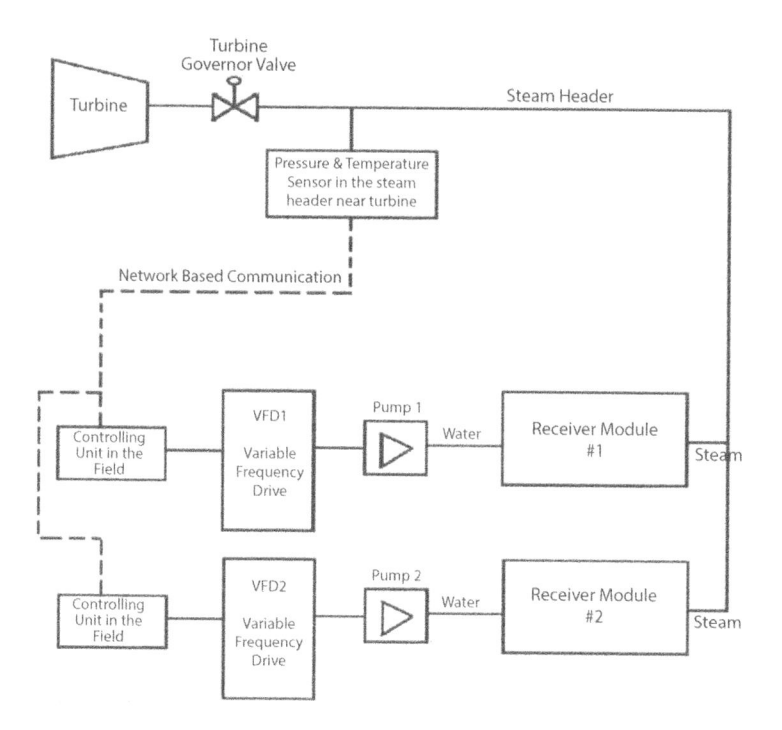

Figure 2.26 PID logic implemented in the control logic.

Table 2.6 Operating data and limit value (Turbine 1).

The machine may be operated with other data only with SIEMENS's consent.			
Power at the coupling	normal	1367.0	kW
	max.	-	kW
Generator power	normal	1305.0	kWe
	max.	-	kWe
Secondary speed	min.	-	rpm
	normal	1500	rpm
	max.	-	rpm
Trip Speed	min.	-	rpm
	normal	1605	rpm
	max.	-	rpm
Gear transmission ratio		9.64	
Live steam pressure	min. (limit value)	38.0	bar
	min. (operation)	-	bar
	normal (operation)	40.0	bar
	max. (operation)	41.5	bar
	max. (limit value)	42.0	bar
Live steam temperature	min. (limit value)	249	°C
	min. (operation)	-	°C
	normal (operation)	252	°C

The project is commissioned in two phases. The first phase was commissioned in November 2016 with 300 nos. of Reflectors commissioned, which is the minimum number required to generate quality steam for steam turbine to operate. The second commissioning was completed in May 2017 with all 770 nos. of Reflector modules commissioned. The project started in full potential from October 2017 with good solar radiation period. Soon after the complete commissioning of the plant with all the sub-systems stabilized to its full potential, the plant achieved record power

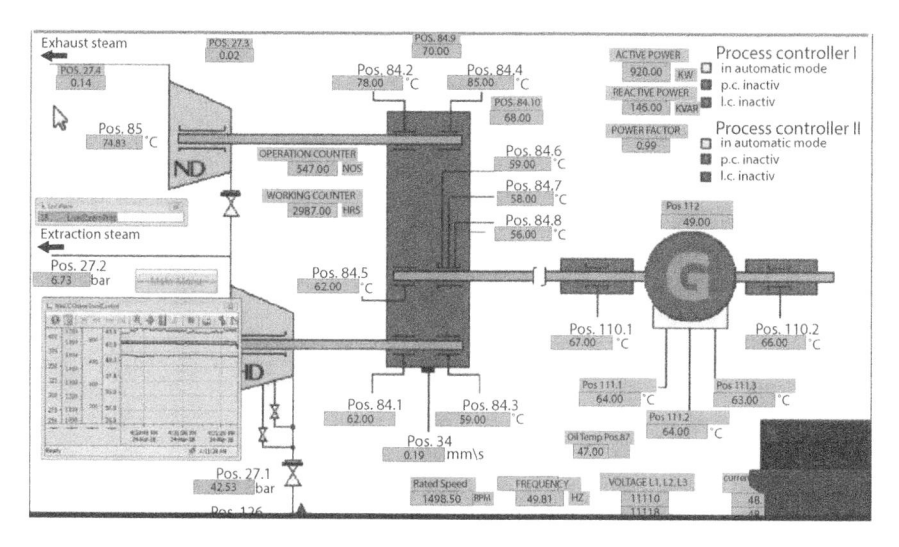

Figure 2.27 Operation parameters and results.

generation for the maximum number of hours of operation without any halt. Another milestone achieved by the plant of the peak output generation. The maximum peak output delivered by the plant till date is 920 KW. The screen shot of the peak output delivered by the plant is shown below. The operation and parameters of the plant is given in Figure 2.27.

2.6 Experimental Data Analysis, Results and Discussions

The thermal output from the reflector is computed by the aperture area of the reflector, the Direct Normal Irradiance (DNI) and the efficiency factor of the reflector. The efficiency factor of the reflector depends on different things, i.e., mirror reflectivity, surface of mirror cleaned, purity and accuracy of mirror, normally it is taken 60% as efficiency factor of reflector surface. Thermal output of 60m^2 reflectors for 2016 is shown as Figure 2.28.

The testing was carried out for thermal behavior of receiver without front glass in the month of May 2011. The Static receiver is charged during the daytime through the solar rays reflection focus from the Paraboloid reflector, shown as Figure 2.29. The Receiver is charged without front glass covering. Observation: Max. Temperature recorded is 400 °C. The front glass cover at receiver opening as shown Figure 2.30. The maximum temperature was recorded to 450 °C and with front glass covering

70 Energy Storage

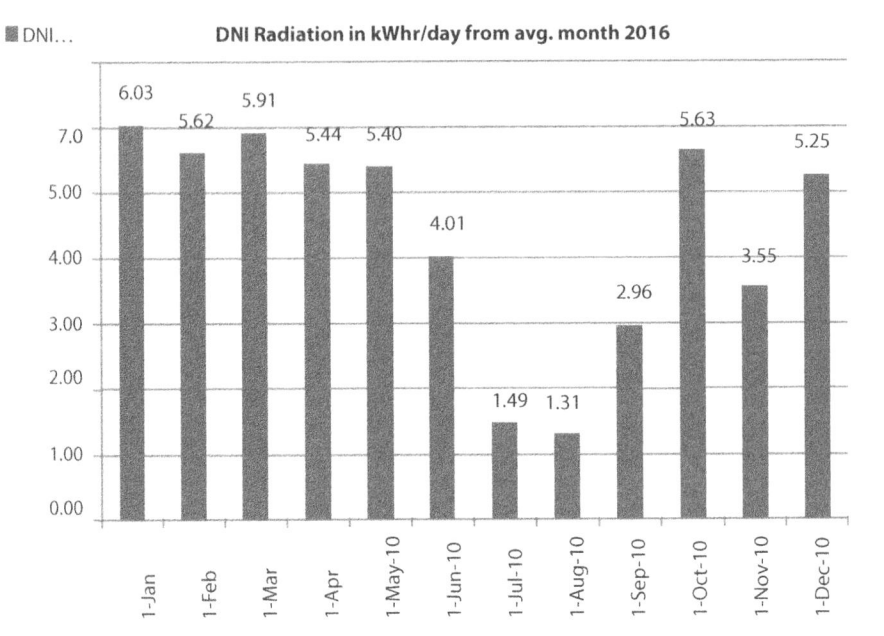

Figure 2.28 Ave. DNI of the location for 2016.

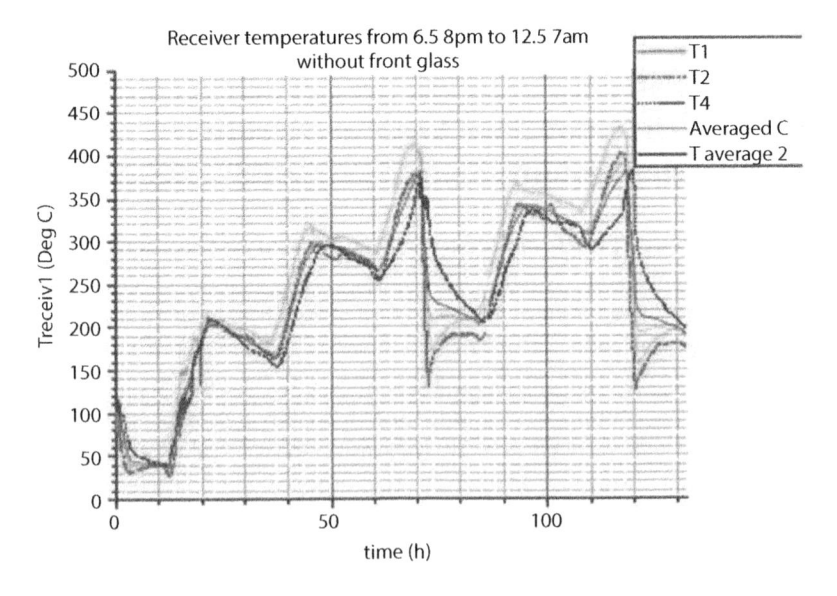

Figure 2.29 Thermal behavior of the receiver without front glass.

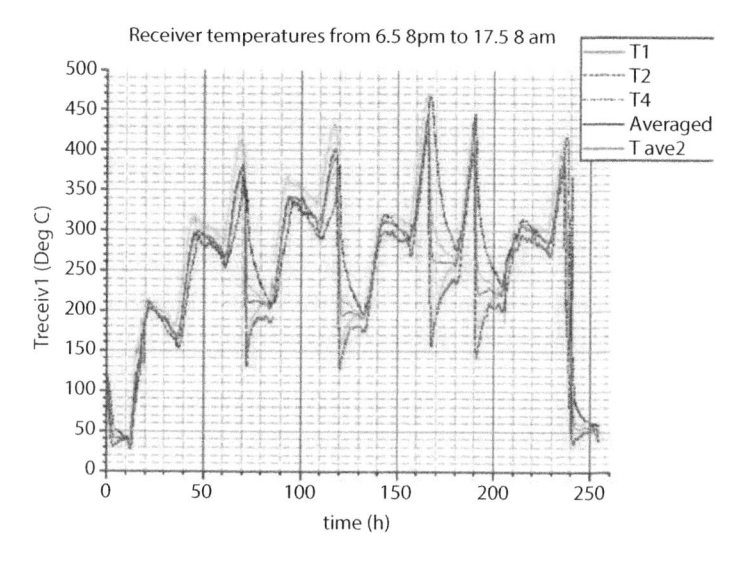

Solar Thermal Power Plant with Thermal Storage

Figure 2.30 Thermal behavior of the cover at receiver opening front glass.

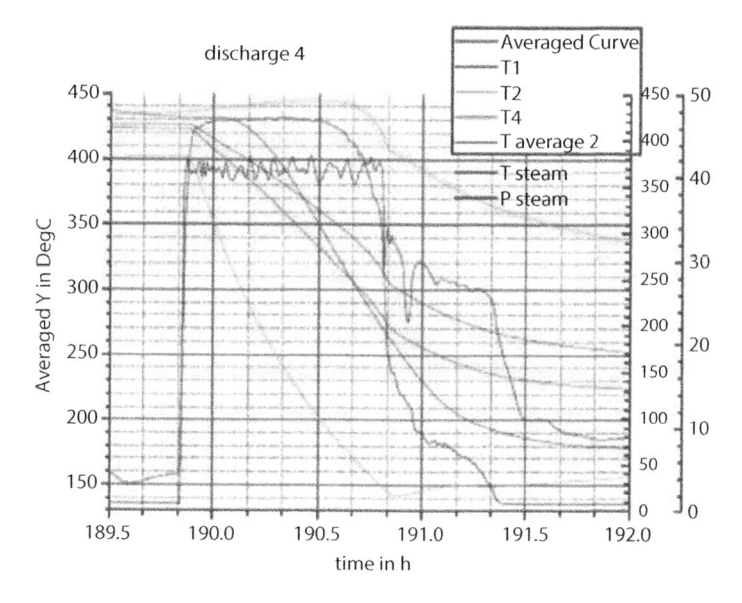

Figure 2.31 Thermal behavior of the Receiver with front glass covering at the receiver opening and with water flow through the heat transfer coil.

at the receiver opening and with water flow through the heat transfer coil, shown in Figure 2.31. The following ovservations were undertaken:

- Flow rate of water through the heat transfer coil is 3.5 litrs/min.
- Max. temperature reached at the Receiver is approximately 450 °C
- Maximum pressure of steam discharged is 42 bar
- Maximum temperature of steam discharged is 430 °C
- Discharging 2/3rd of time superheated steam and 1/3rd of time saturated steam before rapid temperature drop.

Experimental Set up II: Project commenced on 6th September, 2015, then completed on 18th March, 2017. Various observation and measurement have been taken during this period (as shown in Figures 2.32 to 2.35) are as follows:

- Energy storage in form of heat offers a potential pathway for small (local) and large (utility power plants) scale applications. Thermal storage systems provide a unique opportunity to store energy locally in the form of heat that cannot be transported over long distances. Current thermal storage systems are still in their infancy. The most common ones are large, water-heating storage tanks and molten salt-based systems at solar power plants. These systems have been designed based on the economics of water and salt, the heat capacity

Figure 2.32 Filling of salt at core of receiver.

Figure 2.33 Focus of lens at tip of receiver.

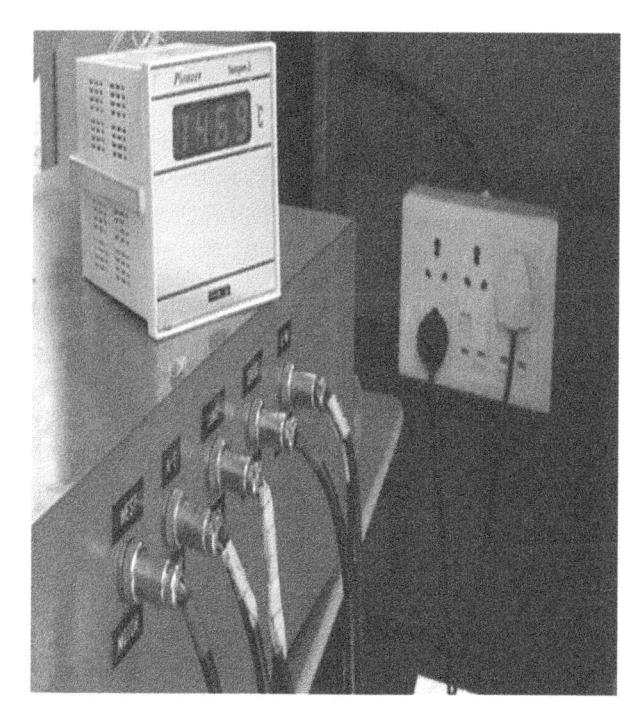

Figure 2.34 Temperature (max.) recorded during field test.

After 6 hours

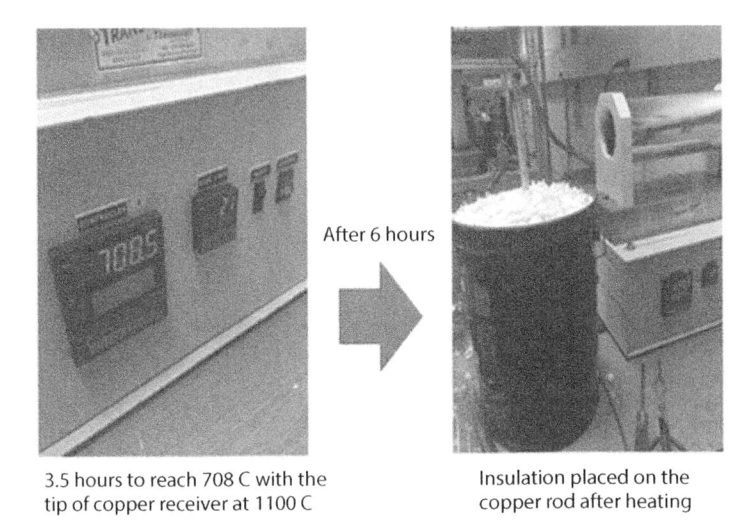

3.5 hours to reach 708 C with the tip of copper receiver at 1100 C

Insulation placed on the copper rod after heating

Figure 2.35 Salt heating and cooling cycle during lab test.

of water, and the latent heat of salts. Research on a large host of sensible heat storage and phase-change materials has been conducted over the past two decades. The materials parameters that are relevant for this application are: melting point, boiling point, vapor pressure, density, heat capacity, thermal conductivity, latent heat of fusion and chemical reactivity.

- While it is intuitive that increasing the temperature of storage could pack in more energy, barriers to the development and deployment of high energy density storage remain, including handling materials at high temperatures, associated systems costs, and operating costs. Thus sensible thermal storage systems are cost prohibitive. Phase change materials (PCM) do provide a viable economical solution for higher energy storage density. However, operation temperatures limit current PCM systems; higher temperatures cause chemical instability and reactivity with containers. Development of an affordable high-density thermal storage system will only be possible by utilizing low-cost earth-abundant thermal storage materials in conjunction with suitable thermally insulating container materials.
- Current heat storage systems utilize either sensible heat storage (i.e., water in storage tanks) or latent heat storage (i.e., phase-change materials such as molten salts). The relatively low operating temperatures of these systems limit

their capacity to store thermal energy; storage systems with higher temperatures would be more economical. In this project, we are developing an affordable high energy density thermal storage system that can store heat at temperature around 1000 ⁰C. The unique aspects of this system are the selection of an alkali halide salt with high melting temperature and a corrosion resistant cheap ceramic container material. The thermal storage unit will be coupled with a high solar concentrator system (1000 – 10,000 x).

- During 3 days of trial run from 11th May to 14th May 2016, we achieved a drop in temperature at the middle of core to the tune of 2 to 2.5 °C per hour in 15 hours from 310 °C to 278 °C. The trial operations will continue for about 6 months to get study steam flow at highest possible temperature and pressure to run the steam turbine for Power generation to the tune of 300W electric (1000W thermal). The temperature achieved at the tip of solar focal point is about 1400 °C. The temperature at the core mid-point was found to be of the order of 310°C, which may be sufficient to generate steam for heat transfer studies.

2.7   Conclusions

Energy security, high efficiency with economy feasibility, and sustainable development with environmental protection are among the most important global topics. In the present era, the growth of population is very fast, and resulting energy demand is also increasing exponentially mainly due to the modern lifestyle, etc. Therefore, renewable-based 24/7 energy solutions have to be invented. Conventional renewable energy generation systems have enormous issues, i.e., uninterrupted supply, energy storage with controlled GHGs emissions. Unlike conventional renewable approach, an innovative passive hybrid approach is the coupling of energy storage system with Concentrated Solar Power (CSP) system. By using solar energy, the hybrid system is able to generate a huge amount of energy. These systems are characterized by various advantages, i.e., appropriate efficiency, no emissions of GHGs with very low operation and maintenance costs, etc. Two experimental set ups with objective to proficient exploitation solar energy and store through solid storage systems to provide the power 24/7. A 1 MWe (3.5 MW thermal) solar power plant with 16 hours thermal storage capacity and A 1 kWe high energy density thermal energy storage for concentrated solar plant were experimented and found satisfactory results as per Indian climatic conditions.

The plant operates on Rankin cycle principle. The Parabolic Reflector concentrates the solar radiation towards the in-house developed, highly efficient cavity receiver. The cavity of the Receiver which is made of monolithic cast iron acts as perfect black body and thus provides excellent thermal storage. The boiler grade coil around the body acts as a heat exchanger which allows for water to exchange heat and convert into steam. It was concluded from various readings that the temperature achieved at the tip of solar focal point was about 1400°C. The temperature at the core midpoint was found to be of the order of 310°C which is sufficient to generate steam for heat transfer studies.

Symbols

kW:	kilo Watts
kWh:	kilo Watt hours
TR:	Tons of Refrigeration
Kg:	kilograms
$:	Dollars
Rs:	Rupees
H:	Hours
m:	Meter
L:	Length (m)
W:	Width (m)
H:	Height (m)
T:	Temperature (°C)
Tin:	Temperature at entry of pipe (°C)
Texit:	Temperature at exit of pipe (°C)
MW:	Mega Watt
GW:	Giga Watt
CO2-e:	Carbon Dioxide-Equivalent
CER:	Certified Emission Reduction
CFD:	Computational Fluid Dynamics
CSP:	Concentrated Solar Power
CL-CSP:	Cross Linear Concentrated Solar Power
DNI:	Direct Normal Irradiance
EAHE:	Earth Air Heat Exchanger
GDP:	Gross Domestic Product

GHGs:	Green House Gases
HTF:	Heat Transfer Fluids
IMD:	India Metrological Department
IPCC:	Intergovernmental Panel on Climate Change
IRR:	Internal Rate of Return
LFR:	Linear Fresnel Reflector
MS:	Molten Salt
NPV:	Net Present Value
PBP:	Pay Back Period
PCM:	Phase Change Material
PDC:	Parabolic Dish Collector
PV:	Photo Voltaic
PSR:	Parabolic Solar Reflector
PTC:	Parabolic Trough Collector
SARA:	Solar Radiation Resource assessment
ST:	Solar Tower
TES:	Thermal Energy Storage

Acknowledgement

The data presented in the chapter is used from the project title "1 MW el. (3.5 MW) solar thermal power plant with 16 hours thermal storage for continuous operation" installed at the site of Abu Road, Rajasthan, India.

References

1. Purohit, I., Purohit, P., 2017. Technical and economic potential of concentrating solar thermal power generation in India. *Renewable and Sustainable Energy Reviews* 78, 648–667.
2. Zhang, J., Chen, T.P., Liu, Y.C., Liu, Z., Yang, H.Y., 2017. Modeling of a selective solar absorber thin film structure based on double TiN_xO_y layer for concentrated solar power applications. *Solar Energy* 142, 33–38.
3. Thirugnanasambandam, M., Iniyan, S., Goic, R., 2010. A review of solar thermal technologies. *Renewable and Sustainable Energy Reviews* 14(1), 312–322.
4. Fernandez-Garcia, A., Zarza, E., Valenzuela, L., 2010. Parabolic-trough solar collectors and their applications. *Renewable and Sustainable Energy Reviews* 14, 1695–1721.
5. Morin, G., Dersch, J., Platzer, W., 2012. Comparison of linear Fresnel and parabolic trough collector power plants. *Solar Energy* 86, 1–12.

6. Gharbi, N., Derbal, H., Bouaichaoui, S., 2011. A comparative study between parabolic trough collector and linear Fresnel reflector technologies. *Energy Procedia* 6, 565–572.
7. Giostri, A., Binotti, M., Silva, P., 2012. Comparison of two linear collectors in solar thermal plants: parabolic trough versus Fresnel. *Journal of Solar Energy Engineering* 135(1), 11001–11010.
8. Ho, C.K., Iverson, B.D., 2014. Review of high-temperature central receiver designs for concentrating solar power. *Renewable and Sustainable Energy Reviews* 29, 835–846.
9. A. Ummadisingu, M.S. Soni, "Concentrating solar power-Technology, potential and policy in India", *Renewable and Sustainable Energy Reviews* 2011;15:5169–75.
10. M. Kamimoto, "Investigation of nitrate salts for solar latent heat storage", *Solar Energy* 1980;24:581–87.
11. Z. Ma, G.C. Glatzmaier, M. Mehos, "Development of solid particle thermal energy storage for concentrating solar power plants that use fluidized bed technology", *Energy Procedia* 2014;49:898–907.
12. X. Luo, J. Wang, M. Dooner, J. Clake, "Overview of current development in electrical energy storage technologies and the application potential in power system operation", *Applied Energy* 2015;137:511–36.
13. H. Sun, X. Luo, J. Wang, "Feasibility study of a hybrid wind turbine system-Integration with compressed air energy storage", *Applied Energy* 2015;137:617–28.
14. S.C. Muller, P.G. Sandner, I.M. Welpe, "Monitoring innovation in electro-chemical energy storage technologies: A patent-based approach", *Applied Energy* 2015;137:537–44.
15. S.K. Soni, M. Pandey, V.N. Bartaria, "Ground coupled heat exchangers: A review and applications", *Renewable and Sustainable Energy Reviews* 2015;47:83–92.
16. S.K. Soni, M. Pandey, V.N. Bartaria, "Hybrid ground coupled heat exchanger systems for space heating/cooling applications: A review", *Renewable and Sustainable Energy Reviews* 2016;60:724–38.
17. U. Herrmann, B. Kelly, H. Price, "Two-tank molten salt storage for parabolic trough solar power plants", *Energy* 2004:29;883–93.
18. S. Relloso, J. Lata, "Molten salt thermal storage: A proven solution to increase plant dispatchability experience in Gemasolar tower plant", *Solar Paces* 2011.
19. N. Breidenbacha, C. Martinb, H. Jockenhöferb, T. Bauerc, "Thermal energy storage in molten alts: Overview of novel concepts and the DLR test facility (TESIS)", *10th International Renewable Energy Storage Conference* Düsseldorf, Germany 2016;15–17.
20. P. G. Bergana, C. J. Greiner, "A new type of large scale thermal energy storage" *Renewable Energy Research Conference* 2014.

21. G. Zanganeh, A. Pedretti, A. Haselbacher, A. Steinfeld, "Design of packed bed thermal energy storage systems forhigh-temperature industrial process heat", *Applied Energy* 2015;137:812–22.
22. Powell K., Rashid K., Ellingwood K., Tuttle J., Iverson B., "Hybrid concentrated solar thermal power systems: A review" *Renewable and Sustainable Energy Reviews* 2017;80:215–237.
23. Lappalainen J., Hakkarainena E., Sihvonenb T, Rodriguez-Garciac M., Alopaeusd V. "Modelling a molten salt thermal energy system – A validation study" *Applied Energy* 2019:233–234; 126–145.
24. Bauer T, Breidenbach N., "Overview of molten salt storage systems and material development for solar thermal power plants" 1–8, https://pdfs.semanticscholar.org.
25. Cristina Prieto C., Osuna R., Fernández A., Cabeza L. "Thermal storage in a MW scale. Molten salt solar thermal pilot facility: Plant description and commissioning experiences" *Renewable Energy* 2016;99:852–866.
26. Cristina Prieto C., Cooper P., Fernández A., Cabeza L. "Review of technology: Thermochemical energy storage for concentrated solar power plants", *Renewable and Sustainable Energy Reviews* 2016;60:909–929.
27. MS Firdaus, SAB Hawa, RI Roberto, MG Scott, "Mirror symmetrical dielectric totally internally reflecting concentrator for building integrated photovoltaic systems", *Applied Energy* 2014;113:32–40.
28. A. Dang, "Concentrators: A review", *Energy Conservation and Management* 1986;26(1):11–26.
29. T. Cooper, F. Dahler, G. Ambrosetti, A. predretti, A Steinfeld, "Performance of compound parabolic concentrators with polygonal apertures", *Solar Energy* 2013;95:308–3018.
30. D. Gadhia, S. Gadhia, Parabolic solar concentrator for cooking, food processing and other applications. http://www.solare-bruecke.org/ SCI conference 2006.
31. T Lee, S Lim, S Shin, DL Sadowski, "Numerical simulation of particulate flow in interconnected porous media for central particle-heating receiver applications", *Solar Energy* 2015;113.20–28.
32. AL Avila-Marin, "Volumetric receivers in solar thermal power plants with central receiver system technology: A review", *Solar Energy* 2011.
33. R Duggal, R Jitle, "Numerical Investigation on Trapezoidal Cavity Receiver Used In LFR with Water Flow in Absorber Tubes", *IOP Conference Material Science and Engineering* 2017;1.
34. B Gobreit, L Amsbeck, R Buck, R Pitz-Pall, "Assessment of a Falling Solid Particle Receiver with Numerical Simulation", *Solar Energy* 2015;115.
35. MJ Busamante, " Developing a solar-bio hybrid energy generation system for self-sustainable waste water treatment", Michigan State University 2016.
36. AA Sinha, Thermal storage possibility in phase changing materials", IJSRD 2017.
37. A Sharma, A Shukla, A Lu, "Low carbon energy supply", *Green Energy Technology* 2018.

38. BP Jelle, A Gustavsen, R Baetens, "The path to the high performance thermal building insulation materials and solutions of tomorrow", *Journal of Building Physics* 2010;34(2):99–123.
39. S Kuravi, DY Goswami, E Stefanakos., "Thermal Energy Storage for Concentrating Solar Power Plants," *Technology and Innovation*, www.researchgate/publication/ 234014902.
40. S Khare, M DellAmico, C Knight, S McGarry, "Selection of material for high temperature sensible energy storage", University of Bath.

3

Efficient Energy Storage Systems for Wind Power Application

Pradeep Kumar Sahu[1]*, Satyaranjan Jena[1] and Umakanta Sahoo[2]

[1]School of Electrical Engineering, KIIT, Deemed to be University, Bhubaneswar, India
[2]National Institute of Solar Energy, GOI, Haryana, India

Abstract

In the present scenario, the development of the wind energy conversion system has increasingly gained attention worldwide to fulfill the global energy demand. The most challenging factor for power system planners and utility operators is the integration of the wind energy to the existing grid due to its intermittency, partial unpredictability, and variability nature in power output. This nature of wind can create problems such as uncertainty in generation, power quality issues, and voltage stability. To overcome all these challenges electrical storage technologies are considered as one of the acceptable and reliable solutions by controlling wind power plant output and providing ancillary services to the power system and therefore enabling increased penetration of wind power in the system. A sole storage unit is not suitable for wind farms due to its restricted capacity. Therefore, the hybrid energy storage system (HESS) technology is more suitable to obtain the expected performance by integrating two or more storage units in various topologies. This chapter focuses on the different power converter topologies used in HESS; interfacing units, power management, and control methods are briefly reviewed here. Finally, the potential of the HESS application in the wind energy system is listed out.

Keywords: Wind power, energy storage devices, hybrid energy storage system, energy management, battery energy storage

---

Corresponding author: pksahu.nitrkl@gmail.com

3.1 Introduction

The prediction of the future consequences of climate change, health hazards due to environmental pollution, energy security, and increasing energy demand, along with the unstable price of the oil market in the past few years, forced the power producer to think about alternate technology such as renewable sources. Simultaneously the restructured electricity market has changed the mode of operation of the power system from vertical to horizontal. This also enables distributed generators like solar, wind, small hydro, biomass, tide, and geothermal to be a vital part of today's electricity market. The reduction of carbon footprint is the main advantage of these sources. Among all the sources, bulk power generation is possible from wind energy after hydropower and the growth has also been remarkable in recent years.

According to statistics given in Figure 3.1, the installed capacity of wind power all over the world by 2019 reached 650.8 GW, according to the World Wind Energy Association (WWEA). In 2019, there was a capacity addition of 59.667 GW. The growth was 10.1% and was considered as the second-strongest wind year. As of now, about 6% of the global electrical power demands are fulfilled by the wind farm [1, 2].

The main drawback of harnessing power from wind energy is the unpredictable wind speed. It may happen the wind farm will be totally shut down due to the variable nature of the wind. This will give rise to security issues like cascade tripping of other traditional generators due to loss of power. In view of this, various Energy storage devices (ESD) are used to supply smooth and continuous power to the system. Energy storage devices (ESD) are utilized by different stakeholders for different types of services.

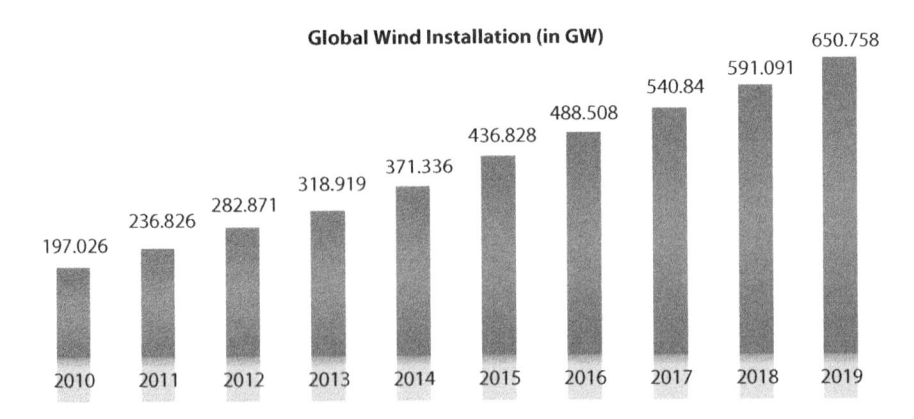

Figure 3.1 Global Wind Installation in GW [2].

Based on the user the values of the service are also different. During the delay of the network expansion storage system is used by transmission and distribution system operators. They are also using this for the management of grid congestion and for providing ancillary services to ensure the stability of the system. Examples of such types of services are black-start capability, Voltage boost up for weak grids, faster frequency control, etc. The power producers integrate the energy storage devices with the traditional generator to optimize power production. The stand-alone energy storage devices are also used for energy arbitrage. Stand-alone renewables (e.g., PV, wind, etc.) power producers utilized the ESDs to mitigate the shortage of power as the renewables are intermittent in nature; on the other hand grid-integrated renewable power producers use energy storage devices during grid congestion to reduce the shortage of electric power and to increase system stability. The ESDs also help the renewable power producers to fulfill their commitment towards the system so as to avoid any type of penalties. ESDs are also used for different types of ancillary service and reactive power compensation even if the wind is not available. Figure 3.2 indicates mainly four types of services for which ESD are used in the power market [3, 4].

Each of the services helps different stakeholders in the various markets for their specific requirements. In this chapter, the services related to wind farm are highlighted. In the current scenario, ESD enables the penetration of wind power to the existing system. Figure 3.3 indicates the different technical aspects which are mitigated by ESD so as to increase the reliability of the system [5]. As the integration of wind power increase, the system reverses and costs consequently increase, while the system reliability and CO_2 reductions decrease [6].

It may not be possible to address all the issues by a single ESD as the power densities and energy densities characteristics for different devices are different. To overcome this issue, suitable combinations of different ESD are used. This combined system is termed as hybrid energy storage system (HESS). In the next section a description of different ESD and HESS is highlighted.

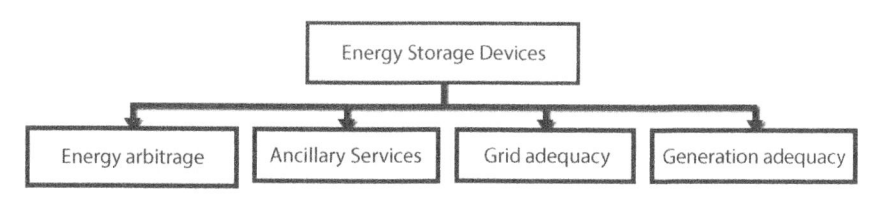

Figure 3.2 Services provided by ESD [3].

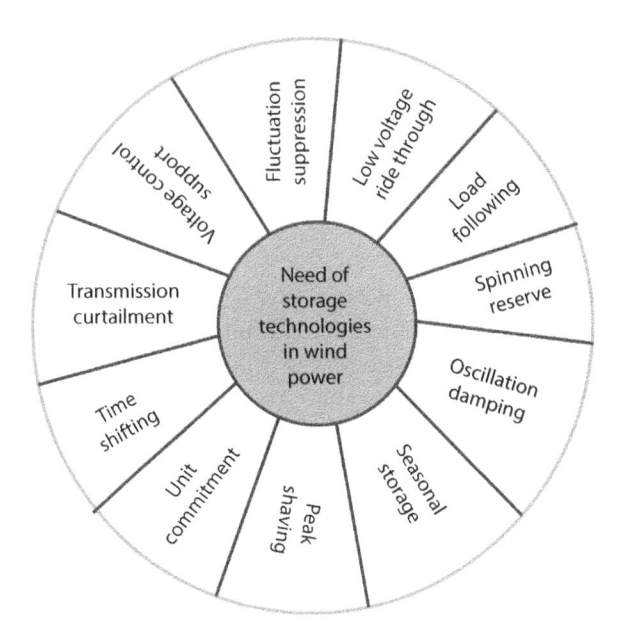

Figure 3.3 Role of storage technologies in wind power [5].

3.2 Energy Storage Devices

The Energy storage technology (EST) is used to store surplus power produced by the wind farm either in the form of electrical energy or some other form like chemical mechanical or thermal. When there is a shortage of power, then the stored energy is utilized to maintain the reliability of the system. The EST is categorized according to different aspects depending on their characteristics such as function, storage duration, etc. In Figure 3.4 a broad classification of the storage devices is highlighted according to the form of how they are storing energy. The first stage of the figure indicates in which form the energy was stored. The second stage indicates the storage medium and the third stage of the figure indicates the technology adopted to store the energy.

3.2.1 Electrical Energy Storage

In this type of storage technology, the excess power is stored in the form of magnetic energy or electrostatic energy. Superconducting magnetic energy storage (SMES) is a device that is used to store electrical energy in terms of the magnetic field and supercapacitors are used to store electrical energy in the form of electrostatic field.

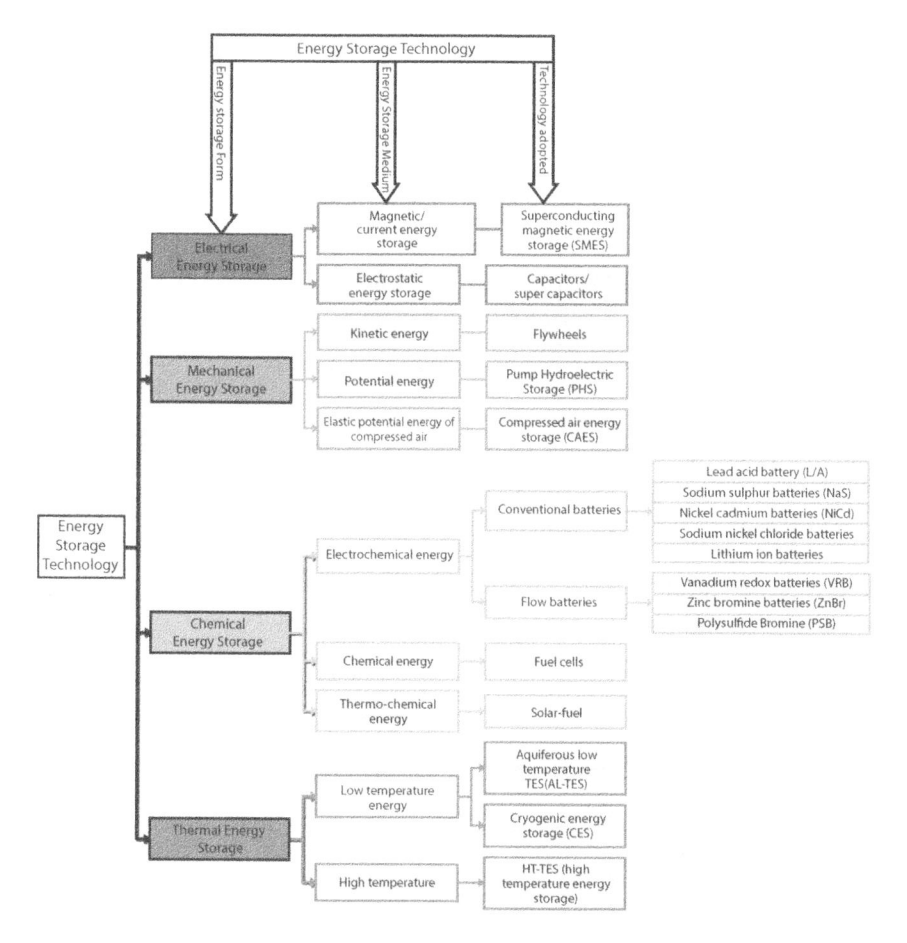

Figure 3.4 Energy storage technology [7].

3.2.1.1 *Superconducting Magnetic Energy Storage (SMES)*

In SMES, the energy is stored by a superconducting coil in the form of a magnetic field. As shown in Figure 3.5, the SMES system consists of a super conducting coil, refrigeration unit, and power conditioning unit. The power conditioning unit supplies DC power to the coil, which stores energy according to the equation given below.

$$E = \frac{1}{2}LI^2 \tag{3.1}$$

Where L is the inductance of the coil. The energy stored depends upon inductance as well as the current passed through it. The resistance of the

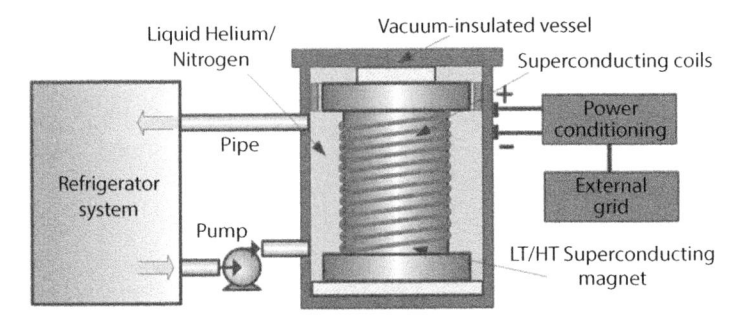

Figure 3.5 Superconducting magnet energy storage [20].

superconducting material is maintained nearly equal to zero to enhance the storage capacity. To achieve such low resistance the superconducting material is cooled cryogenically at a very low temperature.

When there is a shortage of power the energy stored in the superconductor is discharged to the grid through a power conditioning unit. The SMES is characterized by lower energy density, higher power density, and a higher life cycle. One example of superconductor material is niobium-titanium, which is cooled by liquid helium [8, 9]. SMES is used to damp out the variation in wind power as it can discharge within a very short period of time.

3.2.1.2 Supercapacitors

Like any other capacitor, the supercapacitor (SC) consists of two electrodes but there is a electrostatic double-layer capacitance instead of solid dielectric. The double layer is created between an electrolyte and conducting carbon layer. The energy density of the SC can further be increased by using high surface-area materials like activated carbon as shown in Figure 3.6. The SC is also known as ultracapacitors or double-layer capacitors. The SC has much faster response, lower energy density, higher power density (10,000 W/kg) and larger life cycle [10, 11]. The SC is used in wind power

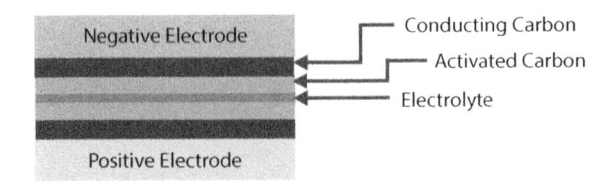

Figure 3.6 Supercapacitor [8].

station to minimize short-term power oscillation and enhance low-voltage ride through capability.

3.2.2 Mechanical Energy Storage

The mechanical storage technology is matured as compared to electrical storage technology as it is used for a longer time. The mechanical storage technology includes flywheel energy storage (FES), pumped hydroelectric storage (PHS), and compressed air energy storage.

3.2.2.1 Flywheel Energy Storage (FES)

Flywheel energy storage (FES) system consists of a rotating cylinder coupled to an electrical machine, as shown in Figure 3.7. The machine acts as a generator during the charging period and stores energy in the flywheel in the form of kinetic energy. During the discharging period, the machine acts as a motor and supply power to the grid [12, 13]. As shown in the figure, the flywheel is kept inside a vacuum to reduce air friction loss during rotation. The energy store in the flywheel is given by

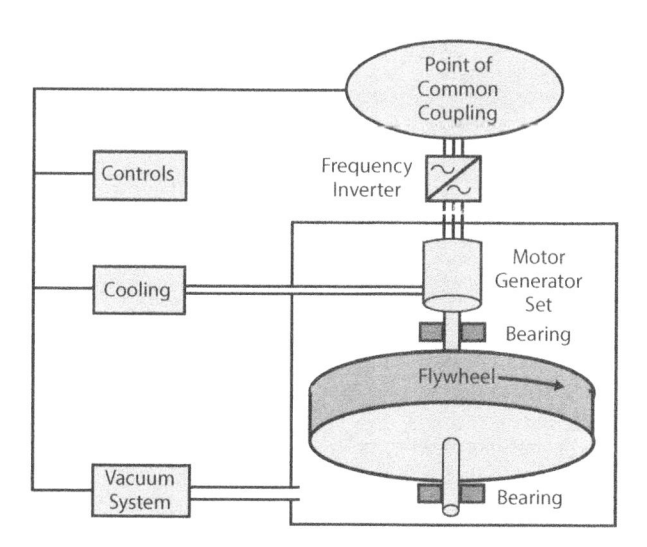

$$E = \frac{1}{2} I \omega^2 \quad (3.2)$$

Figure 3.7 Flywheel energy storage [13].

Where I = moment of inertia and ω is the rotational velocity. The flywheel rotates with more speed to store more energy and slow down to supply the energy to the grid. The FES has high power density but low energy density. Except this FES has a faster response and a longer life period. The FES can be categorized as low-speed FES and high-speed FES which can suppress power oscillation in the wind farm.

3.2.2.2 Pumped Hydroelectric Storage (PHS)

A pumped hydroelectric storage is capable to fulfil the supply-demand gap owing to its larger storage capacity and lesser running cost. A typical PHS system mainly consists of a storage reservoir, pumps, and generation plant. The pump is used to lift the water from the lower reservoir to a higher reservoir during the off-peak hour or when excess wind power is available, which is illustrated in Figure 3.8. The same water again is released from the upper reservoir to the lower reservoir through the generating unit to generate electrical power at the peak hour or at the time of deficit of power.

Generally, mixed flow pumps are used in PHS system as long as the water head is available in the upper reservoir. The PHS system can be categorized as (a) pure or off-stream PHS (b) combined PHS. In the off-stream PHS system uses the water which is previously pumped into the upper reservoir but the combined PHS system utilizes both natural water and pumped water. PHS is generally used to clip and fill gaps in wind power. Although

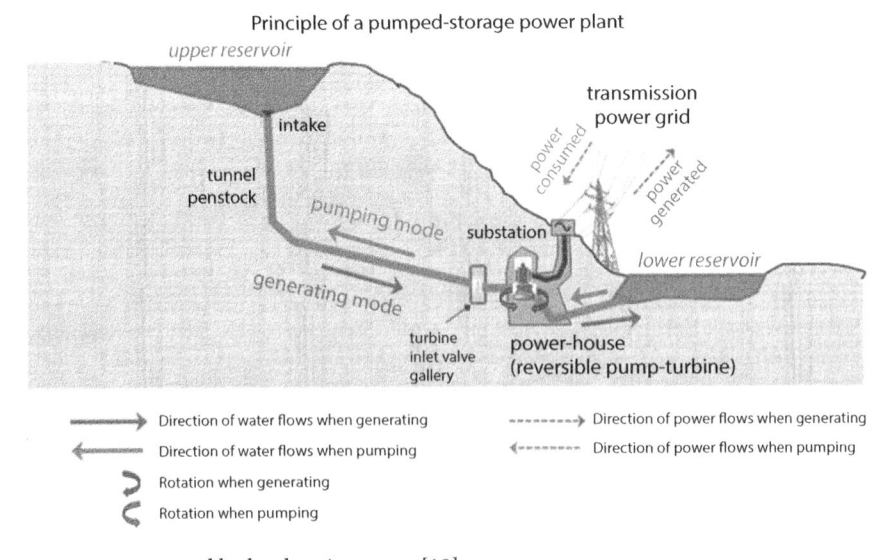

Figure 3.8 Pumped hydroelectric storage [18].

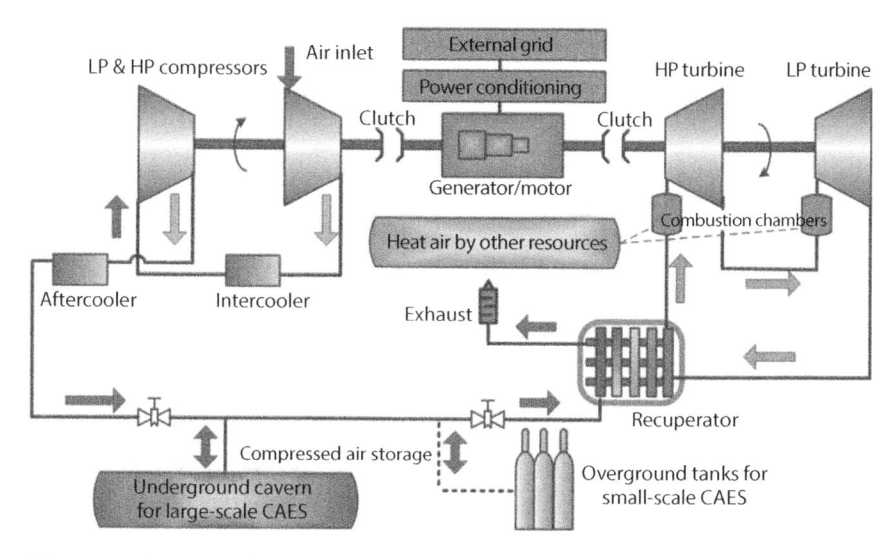

Figure 3.9 Compressed air energy storage [21].

the efficiency is higher by nearly 80% the initial cost and site selection are the major hurdles for the installation of the PHS system [14–18].

3.2.2.3 Compressed Air Energy Storage

The compressed air energy storage (CAES) technology has been used since 1970. The storage capacity of the CAES is high and it is used for large power applications. During the period when power demand is low, the surplus power drives a motor/generator, as shown in Figure 3.9. The motor shaft is coupled to the chain of the compressor. The high pressure compressed air is stored in cravens or overhead tanks. During the period of high-power demand, the compressed high-pressure air is released and heated by means of an external source or from the heat recovered during the compression of air and then the hot air is used to generate electricity with the help of a turbine and generator. To increase the efficiency of the system the waste heat can be recycled with the help of a recuperator. Generally, CAES is used to manage the power output of the wind turbine [19, 20].

3.2.3 Chemical Energy Storage

The chemical energy storage is commonly represented by batteries that have undergone the fastest development with the largest variation. Other types of chemical energy storage are fuel cell and solar fuel cell.

3.2.3.1 Battery Storage System (BSS)

The battery energy storage system is the oldest electrical energy storage technology available in the market. But this is still the most suitable and cost-efficient technology as of now. The energy in the battery is stored in the form of charge due to internal chemical reactions when a voltage is applied across its terminal. The battery discharged the power by reversing the internal reaction. The batteries can be classified into two categories; one is a conventional battery in which two electrodes are present inside the electrolyte and another category is flow-batteries. In flow-batteries two different electrolytes are kept outside the cell in two different tanks. In flow-batteries, the electrolytes are pumped into the cell and separated with the help of a proton exchange membrane for the movement of ion. And the production of electric current. Later it is fed to load via anode and cathode. The different types of flow-batteries available in the market are Vanadium redox batteries (VRB), Zinc bromine batteries (ZnBr), and Polysulfide Bromine batteries (PSB). The widely used conventional batteries are the lead-acid battery (L/A), sodium-sulfur batteries (NaS), Nickel-cadmium batteries (NiCd), Sodium nickel chloride batteries, and lithium-ion batteries. A comparison between different BSS is highlighted in Table 3.1.

3.2.3.2 Fuel Cells

In the market commonly the hydrogen-based fuel cell is very popular as it can be stored and transported very easily. In hydrogen-based fuel cells, by using the surplus power hydrogen is produced by the water electrolysis process in a water electrolyzer as shown in Figure 3.10. At the time of requirement, the chemical energy of hydrogen or hydrogen-based fuel and oxygen from air can be converted to electricity. In the market, six major groups of fuel cells are available according to the selection of fuel and electrolyte. These are Molten Carbonate Fuel Cell, Proton Exchange Membrane Fuel Cell, Alkaline Fuel Cell, Phosphoric Acid Fuel Cell, Solid Oxide, and Direct Methanol Fuel Cell. The process of generation of electricity by the fuel cell is noise-free, pollution-free, and efficient [22].

3.2.3.3 Solar Fuel

Solar fuel is a new technology in which solar fuel is produced by photosynthesis both in natural and artificial processes and thermochemical approaches. By utilizing solar energy, a number of fuels like solar hydrogen, solar chemical heat pipe, and carbon-based fuel can be produced and stored. At the later stage, it can be used to generate electricity [24].

Energy Storage Systems for Wind Power 91

Table 3.1 Battery energy storage system [5, 21–23].

Type of batteries	Name of the batteries	Type of electrolyte	Cathode	Anode	Life span in years	Year of development
Conventional batteries	Lead–Acid Battery(L/A)	Dilute solution of sulfuric acid	lead dioxide	sponge lead	5–15	1859
	Sodium sulphur (NaS)	Beta-alumina solid electrolyte	sulphur	molten liquid sodium	10–15	1970
	Nickel cadmium (NiCd)	aqueous alkali solution	nickel species	cadmium species	10–20	1899
	Sodium nickel chloride batteries	beta-alumina electrolyte	nickel chloride	sodium		1985
	Lithium-ion battery	non-aqueous organic liquid containing dissolved lithium salts	lithium metal oxide	graphitic carbon	14–16	1970

(Continued)

Table 3.1 Battery energy storage system [5, 21–23]. (*Continued*)

Type of batteries	Name of the batteries	Type of electrolyte	Cathode	Anode	Life span in years	Year of development
Flow Battery	Vanadium Redox Flow Battery (VRB)	aqueous acidic vanadium sulfate (V2+/V3+ are used as electrolytes in one tank and V4+/V5+ electrolytes in another tank)	carbon felt electrodes	carbon felt electrodes	5–10	1980
	Zinc bromine batteries (ZnBr)	aqueous electrolyte solutions of Zinc element in one tank and aqueous electrolyte solutions of Br element in another tank	carbon-plastic composite electrodes	carbon-plastic composite electrodes	5–10	1970
	Polysulfide Bromine batteries	sodium bromide in one tank and sodium polysulphide in another tank	carbon felt electrodes	carbon felt electrodes	10	1983

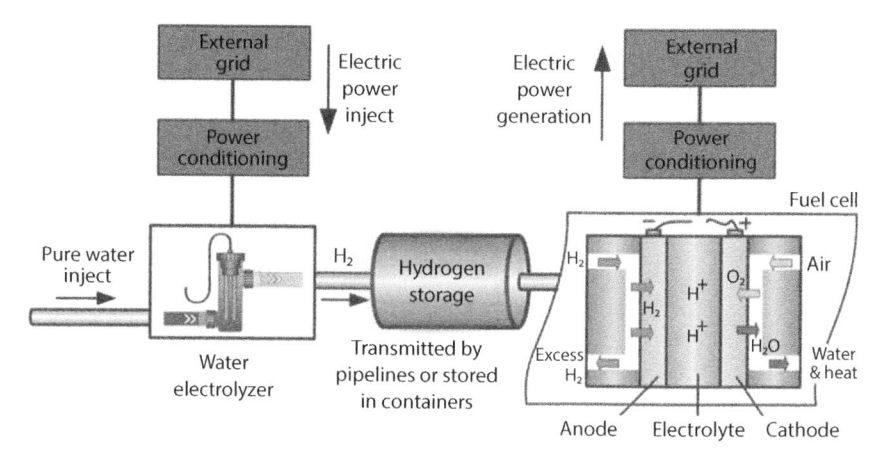

Figure 3.10 Fuel cell [21].

3.2.4 Thermal Energy Storage

In a thermal energy storage (TES) system energy can be stored without any hazard in a large quantity. The self-discharge losses of TES systems are also very less. The TES consist of mainly a storage medium in the form of a tank or reservoir, a refrigeration system, pump pies, and a control structure. In TES different technologies are used to store the available heat energy in insulated repositories using the different concepts. TES can be divided into two groups, the low-temperature TES (LT-TES) and high-temperature TES (HT-TES). The LT-TES further consists of cryogenic energy storage and aquiferous LT-TES. Similarly, HT-TES includes sensible heat TES latent heat of fusion TES and concrete thermal storage. TES is widely used for the generation of electrical energy from the heat engine and load shifting [22].

3.3 Hybrid Energy Storage System (HESS)

The maturity level of different ESS is shown in Figure 3.11. The ESS technology is categorized on the basis of developing stage to fully mature stage.

The energy storage mechanism has two major components, energy density and power density. The ESD like fuel cells and batteries has high energy density and low power density. These types of storage units have a slow dynamic response and hence face power regulation issues in the network. However, storage units like flywheel or supercapacitors are able to provide high power demand. This may reduce the lifespan of the ESD [24]. Both characteristics (energy density and power density) cannot be achicved by

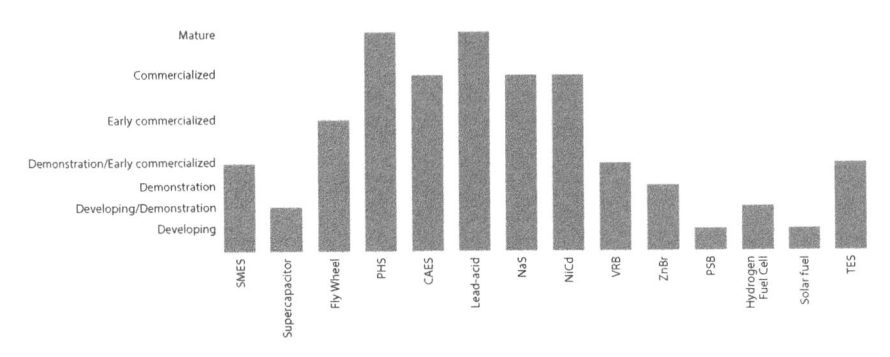

Figure 3.11 Maturity level of storage technology [21].

the presently available storage technologies. Therefore, steady-state and dynamic performance can be enhanced by using HESS technologies. For stable and economical operations, wind generators require proper storage technology as well as an appropriate control method.

Generally, wind farms face challenges like inconsistency nature, stability problems, power quality issues, and fluctuation of voltage and frequency of the system. In the wind farms, the replacement and operation cost of storage units is enhanced considerably due to the frequent and irregular charging and discharging process. The lifespan of the ESDs is also reduced due to the improper charging and discharging operation. Keeping the above issues in mind, HESS technology is more suitable for wind power applications. The technology and economic aspects of HESS in the wind farms are briefly reported on. The HESS can be formed by the hybridization of various energy storage units. Usually, HESS technology consists of two components like high-energy storage (HES) unit and high-power storage (HPS) units. The long-term energy demand is supplied by the former storage unit whereas the later storage unit meets the short-term peak and transient power. The HESS technology has several advantages in wind energy applications like better efficiency, low cost, and improved lifespan of the storage units. The hybridization among storage units can be formed by taking different storage technologies with their characteristics. The hybridization of various storage technologies used for wind farms is shown in Figure 3.12 [25]. The different combination of ESS are more suitable for wind power applications. The proper HESS configurations depend on various factors like hybridization targets, cost, space required, and geo-location.

Four factors should be kept in mind while designing the HESS technology for wind farms like the capacity of types of storage unit, power converter topology, energy management, and the control strategy which

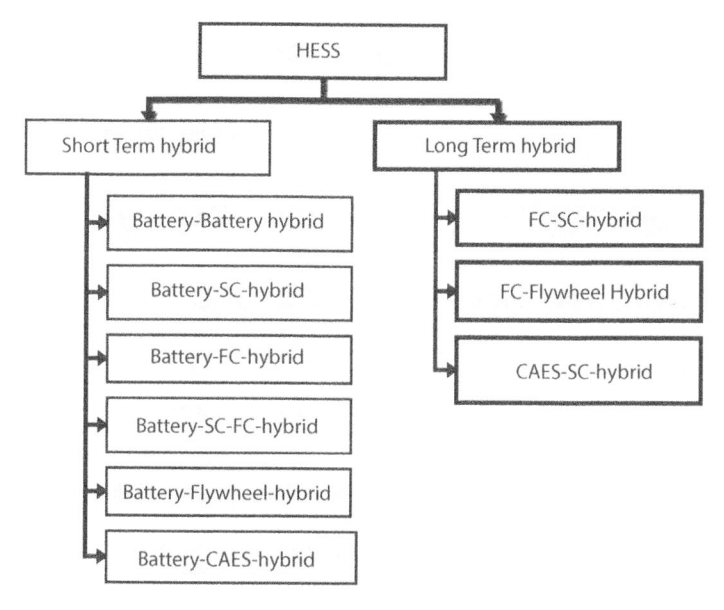

Figure 3.12 Hybrid configuration [25].

should be dealt with carefully. The above aspects are highlighted in the next section [25].

3.4 Power Converter Topologies for Hybrid Energy Storage

Battery-based hybrid energy storage systems (HESS) can be integrated with the wind generator through various topologies. High power storage (HPS) and high energy storage units can be interconnected through the different topologies [26–28]. A brief comparison among HESS topologies has been thoroughly discussed in [29]. The converters used for the integration in wind farms are categorized as passive, semi-active, and active, which is shown in Figure 3.13.

3.4.1 Passive Topology

Here, two storage units are simply interconnected, so that it makes a simple, efficient, and cost-effective topology [30–33]. The power management among HES and the HPS units is generally regulated by its internal resistance and its voltage-current relation curve. In this topology, the terminal

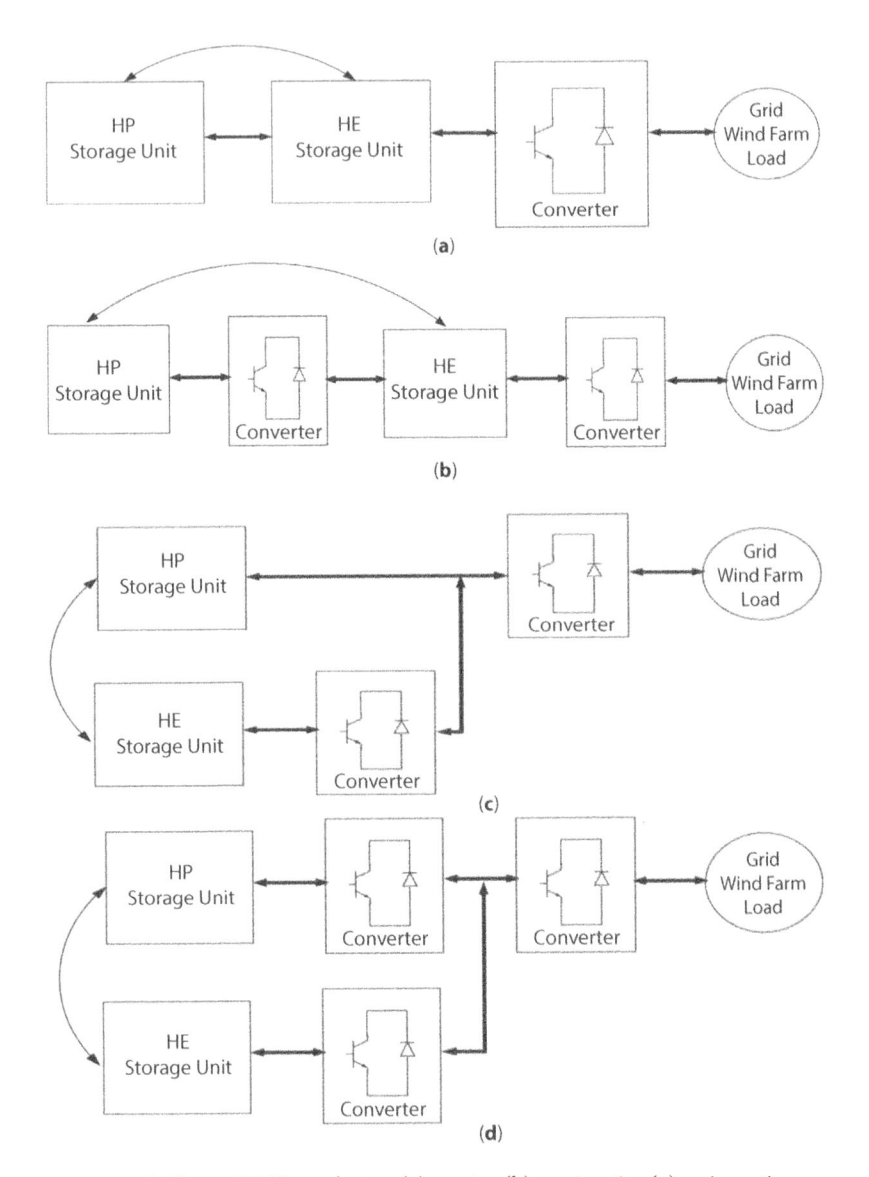

Figure 3.13 Different HESS topologies. (a) passive (b) semi-active (c) series active (d) parallel active. (*Continued*)

ENERGY STORAGE SYSTEMS FOR WIND POWER 97

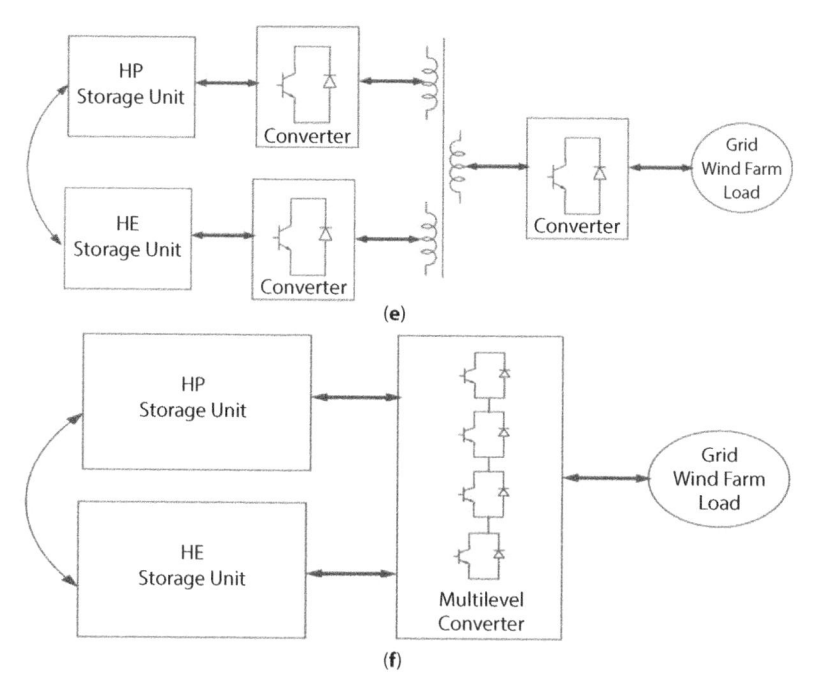

Figure 3.13 (Continued) Different HESS topologies. (e) isolated active, and (f) multilevel topology.

voltage of the storage units is not controlled. Therefore, the HPS's accessible energy is limited and it behaves like a low pass filter for the HESS.

3.4.2 Semi-Active Topology

In this topology, the terminal of one storage unit is connected with a power converter whereas the other storage unit is inserted directly at the dc-bus. So, the application of a power converter needs additional space and is also more costly. But this topology provides better controllability and power dispatch ability. Various semi-active topologies for HESS are briefly reported in [34]. The application of the additional power converter in this configuration confirms a wide operating range of the HESS [35].

3.4.3 Active Topology

In this topology of HESS, two or more storage units are connected to the wind farm through the separate power converter; as a result the system becomes complex, with more losses and higher cost. However, this topology has some advantages like better controllability of each storage unit through the individual power converter. The active HESS topologies can be controlled by using sophisticated control techniques [36]. The parallel active HESS configuration includes two power converters for power management among the storage units which is shown in Figure 3.13 (d). But in conventional parallel configurations, the power management is through the common dc bus. Moreover, the two parallel converters affect negatively the system efficiency. This problem can be overcome by the modified topology which is reported in [37]. The modified topology possesses advantages like small size dc-bus capacitor and better efficiency, as well as reconfiguration facility as compared to conventional active topologies.

The HESS also includes multilevel power converter as the power conditioning unit which is reported in [38–43]. The use of a multilevel power converter in a wind farm makes the system more reliable and improves power quality. The cost and control complexity can be reduced by connecting several storage units in a single power converter. But the multilevel power converter consists of more semiconductor power switches and capacitor which makes the controller more complex. The HESS is also integrated with the wind farm by using Isolated multiport power converter topology which is briefly reported in [44–46]. Each topology has its own merits and demerits, but currently, the active topology is commonly used due to its high capability to handle a large amount of power in the network. The different power converter topologies are also used for the HESS integration with wind farms which is briefly discussed in [47, 48]. Several factors like cost, flexibility, efficiency, and reliability are taken into account to select a power converter topology for wind energy system.

3.4.4 Comparison of Different Topologies

The energy management technique of HESS directly depends on the power converter topology. One cannot control the power of storage units directly. However, the terminal voltage of one storage unit will be the same as the dc-bus voltage, whereas the output power of the other storage unit is uncontrollable in the semi-active topology. In the case of active topology, both the input and output power of both the energy storage can be controlled by using a rational controller by compromising its efficiency. Several factors like cost, flexibility, efficiency, and reliability are taken into account to select appropriate power converter topology for a wind energy system. The comparison among various HESS operational aspects are given in Figure 3.14. The passive configuration has a cost-effective and simple structure but non-controllable. The active HESS topology is suitable for the wind

■ Active Topology   ■ Semi Active Topology   ■ Passive Topology

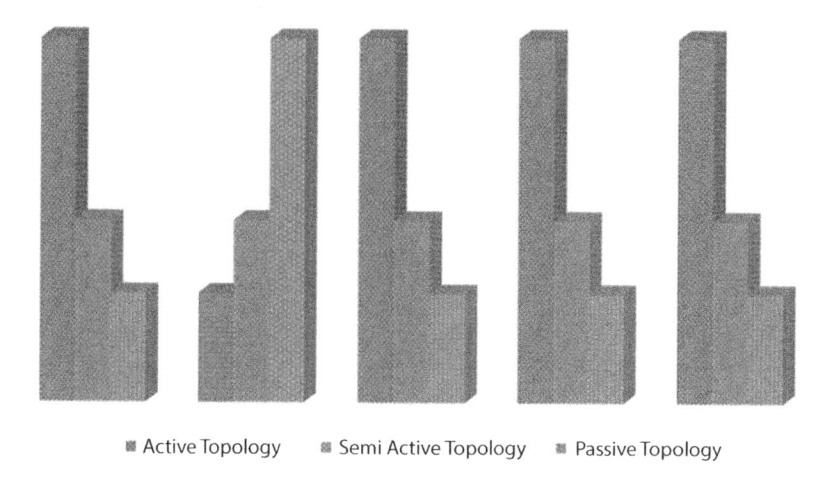

Figure 3.14 Comparison of different topology.

energy system due to its better controllability and flexibility to reconfigure with higher cost and more complexity. However, a semi-active structure includes lower costs with restricted controllability.

3.5 HESS Energy Management and Control

The energy management of HESS for wind energy system depends on designing and implementing an appropriate control scheme. The various factors should be kept in mind for choosing the proper control technique for HESS. The control technique for HESS can be selected on the basis of the types of system (wind farm, microgrid, grid-connected or not), controller cost, controller response time, and configuration of HESS. The safe and stable HESS operation can be achieved by using a proper power-sharing technique. The HESS control scheme, as well as energy management of HESS, are categorized as (i) the energy management unit, and (ii) the underlying control unit. The brief clarification of the HESS system is shown in Figure 3.15.

3.5.1 HESS Control Schemes

Power management in HESS is a prime challenge for wind farms. Different control schemes are reported for power management in HESS. The power allocation techniques are classified into two types: (i) classical control

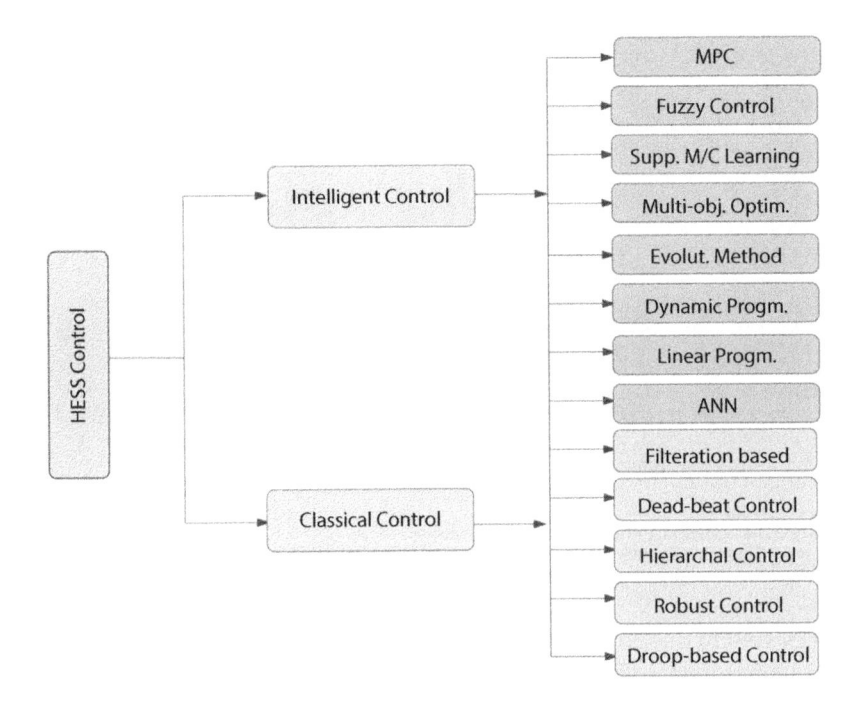

Figure 3.15 Energy management classification of HESS.

schemes, and (ii) intelligent control schemes. Different control techniques used for HESS are discussed in this section.

3.5.1.1 Classical Control Scheme

This control scheme is further classified into three parts: droop controllers (DC), a rules-based controller (RC), and filter-based controller (FC). A mathematical model can be developed for designing these controllers.

The power management between the various storage units of HESS can be achieved by the active coordination control methods. Again, this coordination controller is divided into three classes: centralized, decentralized, and distributed control.

Among these control schemes, the decentralized control scheme is suitable for wind farms as the other two control schemes face problems like communication failure. One of the efficient decentralized control schemes is the droop controller (DC). A brief discussion about different DC used for HESS is reported in [49]. Advanced droop control schemes like integral DC [50], adaptive DC [51], virtual resistance DC [52], virtual capacitance

DC [53], virtual impedance DC [54], filter-based DC [55], and voltage recovery DC [56] are commonly used for wind farm applications.

In RC, power management among storage units can be achieved by using some predefined rules. This controller is further classified, such as: state machine, power follower, and thermostat methods [57]. These types of control schemes are simple in design and operation. RC is an efficient technique for the energy management of HESS. However, this method is more sensitive to parameters fluctuation.

In the FC scheme, the total power of the system is dissolved into low- and high-frequency components with the help of a filter. Here, the filter parameters decide the reference power for storage units of HESS. The high-pass filter and low-pass filters with different ramp rates are used for arranging the reference current among storage units and resolve the net power into high- and low-frequency components.

In the filter methods, the voltage across the dc-bus is regulated for allocating power among the storage units which is shown in Figure 3.16. This method of control is easy to implement and also less expensive. However, this control scheme is not effective for more wind generators and storage units in a microgrid because power management is not accurate.

The deadbeat controllers [58] are also commonly used for HESS to enhance transient performance. This method is more efficient than the conventional PI controller because it is always operated at an optimal duty cycle. The stress on battery units can be reduced; as a result, the lifetime of the battery is extended. This control scheme can be easily implemented with reduced cost and less number of the sensor. But, the performance of the control depends on the mathematical modeling of the system. When the system's exact model is not possible and fluctuation in the parameter of the system, it could not operate properly. Another advanced controller called Hierarchical controller is also commonly used for HESS [59]. This control method is robust against load fluctuation and variation in wind

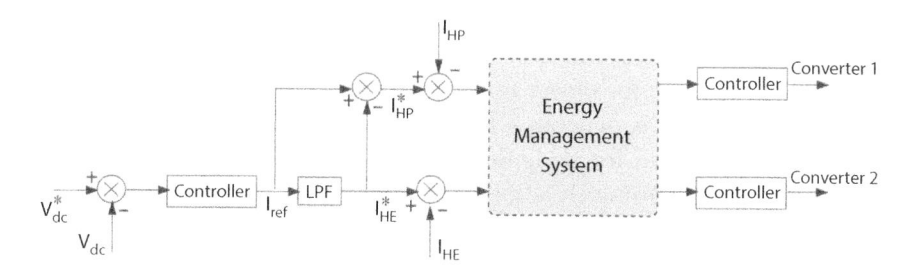

Figure 3.16 Filter-based control scheme for HESS.

power by using a stochastic control scheme. The hierarchical controller uses high precision intraday forecast data. This control method is more stable and more economical for HESS [59]. It comprises hour-ahead and real-time scheduling. The control scheme has a higher life period of the battery units used in the wind farms. Both centralized and distributed control configuration of this controller improves the reliability of the wind farm.

3.5.1.2 Intelligent Control Schemes

An intelligent control scheme works on the basis of optimization methods that are suitable for a complex system. These optimization approaches are Fuzzy logic control, artificial neural network, dynamic programming, linear programming, genetic algorithm, and model predictive control. Any complex system can be efficiently controlled by using a Fuzzy logic controller. Simple in design is the main feature of this control scheme. Unlike the classical control method, the exact mathematical modeling of the system is not needed for this control scheme. The power-management in HESS based on the Fuzzy logic control scheme is briefly reported in [60, 61]. The protection of the battery from overcharging and discharging can be achieved by a dedicated mode of this control scheme. The peak current of the battery can be limited by using PSO optimized algorithm [61].

The energy management among the storage units can also be achieved by using an adaptive Fuzzy logic control scheme [62]. This controller has properties like easy to implement, better efficiency, and less current fluctuation for storage units.

The energy-management for HESS is presented by using an advanced control technique; model predictive control scheme (MPC). The reference for the storage units and power-sharing among them can be obtained by using this control scheme [63]. The system consists of a battery/fuel cell/supercapacitor controlled by MPC for power allocating among the HESS units [63]. The HESS faces various constraints like the variation in load, shut-down cycles, and start-up cycle which is mitigated by the MPC scheme. A reference for power management among the battery, fuel cell, and supercapacitor is generated for this controller. Different challenges need to be addressed for designing the MPC controller like operational restriction, downgrading situation, and overall cost. The MPC controller based on duty cycle control of power converter is given in Figure 3.17. This controller determines the optimal duty cycle of the dc-dc converter which will control the power provided by the battery and supercapacitor.

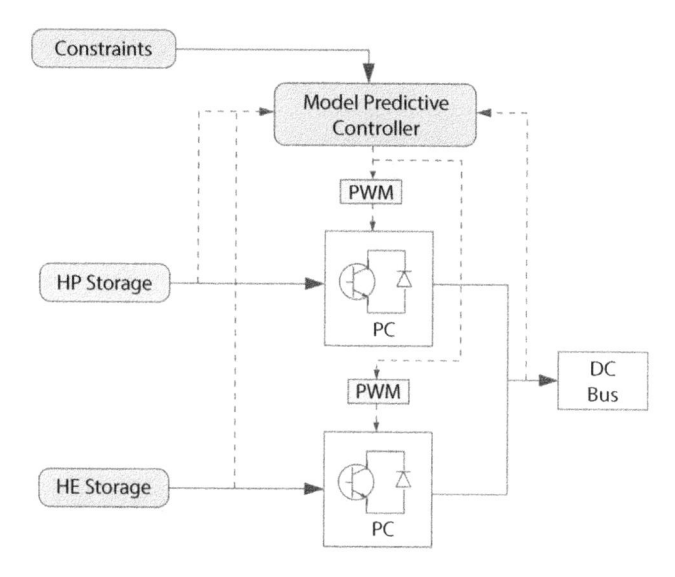

Figure 3.17 MPC control scheme for HESS.

The artificial neural network (ANN) control scheme does not need the accurate mathematical modeling of the system. These controllers are suitable for wind energy applications due to their pattern recognition facilities. This algorithm is used for forecasting the velocity of the wind for the wind turbine. It utilizes the historical dates of the wind farm location for tuning the control parameters. The grid-connected hybrid system consists of wind farm, PV plant, fuel cell, and battery bank which are controlled by this controller as reported in [64]. ANN-based control scheme extracts maximum power from the RE sources by using the MPPT technique and regulates the energy interchanged among the utility grid and the power converter.

3.5.2 Comparison of Different Control Schemes

There are no unique methods adopted for selecting the control schemes for the energy management of HESS. Several factors should be kept in mind before applying a particular control scheme in wind farms like the role of the controller, the cost of the controller, response time, and types of storage units used in the system. All the HESS controllers discussed in the previous section have their own merits and demerits. Among all the above control schemes, classical controllers have better performance due to reduced computational burden. But, this method is not suitable for the systems which are difficult to develop the mathematical model. On the other hand, intelligent controllers are flexible and various cost functions may be taken, but it has a disadvantage like a large computational task.

3.6 Applications of the Storage Technologies in Wind Power

The application of the energy storage system (ESS) in wind power is briefly discussed here. Various operational issues of the wind power farms with their solutions and their effect on the electrical network have been reported in [65, 66]. Moreover, the role of ESS in each issue is briefly discussed in this section.

3.6.1 Power Fluctuation Mitigation

The inconstancy in output power (duration up to a minute) of the wind generator may lead the fluctuation in frequency as well as output voltage, particularly in the islanded system. This phenomenon may cause power quality issues in the network [67]. The storage systems are commonly used in WPS to suppress the output power variations affect. For uninterrupted operation, the storage systems should have properties like high ramp power rates and longer life cycle Therefore, storage technologies like batteries, supercapacitors, flywheels, and SMEs are suitable for the WPS.

A commonly used technique to suppress any variation in the power of the DFIG is to incorporate a storage unit in the dc-link of the back-to-back converters. The power injected into utility grid can be optimized by connecting these storage units with the proper control techniques. A supercapacitor is employed with the dc-link through a bidirectional dc-dc converter topology which is reported in [68]. A two-layer control scheme is employed to manage the power injection into the grid. The first layer, which consists of supervisory control, is used for coordinating the set point of each wind farm, whereas the lower layer uses a vector controller for the converters connected with each wind generator.

Other storage units like flywheel are also commonly connected across the dc-link of the wind generator. The working principle of Flywheel technology is based on the energy management of a motor/generator, hence these types of study require knowledge of speed control of electrical drives. The flywheel is driven by a vector controlled induction motor drive which is reported in [69]. Moreover, model reference-based adaptive control schemes are used for flux weakening control and estimation of the speed of the flywheel [70]. In the wind farm, permanent magnet synchronous machine, induction, and switched reluctance machines are analyzed for the flywheel units [71]. The power variation of the wind generator will lead to the power quality problem, particularly in a stand-alone system. The power quality of the stand-alone wind farm can be enhanced by connecting the hybrid storage system which is a combination of supercapacitor, batteries, or flywheels [72].

3.6.2 Low Voltage Ride Through (LVRT)

The wind farm with the voltage control loop is connected with the utility grid at the point of common coupling. The function of the voltage control method is to protect the wind farm at the time of voltage dips. Here, the wind farm takes grid codes to tolerate the voltage dips within a certain period. This phenomenon is called LVRT. The reactive power supplied to the utility grid can be adjusted by using the power converter of the wind farm [73]. Hence, HESS is not required for the above condition. However, it will secure the dc-link against overvoltage conditions.

The proper HESS is selected for suppressing the LVRT phenomenon. It requires a high ramp-up for the power converters. Hence, storage units like a battery, short-duration storage units such as flywheel or supercapacitors are commonly used for wind farms. The storage unit connected across the dc-link of the back-to-back converter is reported in [73]. Several studies confirm that storage units improve ride-through ability by applying the FLC method.

3.6.3 Voltage Control Support

Induction machines used in wind farms generally utilize more reactive power. The system voltage can be maintained to a predefined value by regulating the reactive power of the network. Several methods are reported in [74] for reactive power compensation of wind farms. The power generated by the DFIG or synchronous generator delivers to the electrical circuit through converters. These power converters can easily regulate by applying reactive power and voltage control methods. Moreover, the dynamic voltage control could be enhanced by integrating storage units in wind farms.

Hence, storage units like a battery, short-duration storage units such as flywheel or supercapacitors are commonly used for wind farms due to its better ramp rate. The storage units are integrated with the wind farm through power converters which are used for the reactive and active power of the network. The DSTATCOM combined with the storage units is used in the wind energy system to minimize the voltage stability issues [75]. Here, the storage is connected across the dc-link to interchange both the power. The STATCOM used in a wind farm has advantages like reactive power compensation, harmonic reduction as well as enhance dynamic performance under load fluctuations. This confirms better system efficiency.

3.6.4 Oscillation Damping

Here, storage units are connected with a wind generator to supply energy to the network for a duration of ten hours [76]. The power supplied by the wind generator cannot meet the required load demand due to the intermittent character of wind speed. This phenomenon causes different operational problems like frequency as well as voltage fluctuation due to a mismatch of power supply and demand. This mismatch of power also leads to the economic challenge like financial penalties. Therefore, storage units are integrated with the wind farm to supply additional power for a long time. Storage units like a battery, short-duration storage units such as flywheel, HESS, or supercapacitors are commonly used for wind energy applications.

The wind farm installed in Japan consists of wind generators of capacity 52 MW and a storage battery bank of rating 35 MW. The storage units require appropriate energy management techniques in terms of economic issues and technical problems like scarcity or excess power of storage units. The economic analysis is investigated by using an energy management control scheme which reduces the use of storage units during peak hours when the cost is quite high.

Sizing and the control scheme for the battery used in wind applications are reported in [77]. This paper focuses on the techno-economic analysis of storage-based wind farm. The reliability of the system can be improved by using appropriate control techniques and also leads to a reduced storage requirement. Two storage units of the capacity of 40 and 36 MWh are employed to enhance the power generated by a 100 MW wind farm [77]. In this paper, an optimization algorithm is used for addressing issues like power demand, energy storage requirements, economic aspects, transmission facilities.

3.6.5 Peak Shaving

Peak shaving application for wind farm operates in the time range of one to ten hours. Here, the energy storage units reserve the energy during the off-peak conditions and supply the energy to the network during peak-hour. This phenomenon minimizes the peak and valley profile of the load curve.

The storage units like batteries, HESS, compressed air storage systems, and pumped hydro storage systems are suitable for peak shaving operations. Numerous research studies have been conducted on techno-economical analysis for the battery to supply power in peak-hour and reserve the energy in an off-peak hour. The use of sodium-sulfur batteries for peak shaving applications are reported in [78]. This paper focused on the merits of battery usage in the real field and power conditioning units used for integrating with wind farms. The brief techno-economical feasibility is also discussed in [79]. In the Spanish energy market, the selling of storage energy is kept at a fixed price for the economical operation of the battery. To ensure the economic operation of battery in peak saving application, the operators may provide subsidies to discourage the use of traditional fuel resources. The cost of storing energy can further decrease by using a regenerative fuel cell system [80]. It shows that the energy cost of the regenerative full cell is 7 to 8 times more than the hydrogen-based storage system.

In an isolated system, the pumped hydro storage system is the most suitable candidate wind energy system as it has a better penetration range. The application of pumped hydro storage with wind farms is thoroughly reported in [81]. In this paper, the pumped hydro-wind are supplying power at peak hour and a detailed techno-economic feasibility study is carried out. The research indicates tremendous performance in terms of efficiency and economy. The pumped hydro storage with wind farms not only increases the penetration level by 9% but also shows a considerable decrease in CO_2 emission. The spinning reserve facility of this storage system can be improved by adding additional technology [81].

3.6.6 Spinning Reserve

The unaccustomed space of storage units which could be altered by the researcher and is provided synchronizing with the network devices capable of affecting the active power of the system. The spinning reserve can be achieved by activating the secondary and tertiary reserve. In this scenario, the active power of the system can be regulated by the wind generator to facilitate the frequency support for approximately thirty minutes.

Numerous storage units like a battery, superconducting magnetic energy storage, flywheel, pumped hydro storage unit or HESS are commonly used in wind energy applications. The wind farm facilitates the spinning reserve ability by using the battery storage system which is briefly reported in [82]. In a stand-alone hydro-wind-gas power system, the role of the battery for spinning reserve capability is briefly investigated in [82]. To ensure the optimal economic operation in wind application, the sizing of the battery as well as energy management while satisfying the reserve capability has been discussed in this paper. Various optimization techniques are reported for using the battery.

The spinning reserve capability can also be achieved by using a flow battery. Due to the properties like overload handling capability and short-time response, this storage system is the most suitable candidate for availing spinning reserve function [83].

3.6.7 Time Shifting

In this application, the energy from the storage units has to supply to the islanded system for a time range from five to twelve hours. Here, the energy from the wind farms during off-peak hours is required to store in the storage units. The power stored during this period is quite cheap and the power can be dispatched to the network during peak hour. This leads to the discouragement of using conventional peak power plants like diesel generator. The storage units like pumped hydro storage units, superconducting magnetic energy storage systems, flow batteries, or hydrogen-based storage units are commonly used for wind power applications.

The operation of the hybrid system consists of wind farms and bulk storage units is briefly explained in [84]. In this study, the battery is charged continuously for a long period (nearly 12 h) during off-peak power and supplied the regulated power to the network during peak-load hour. In this manner, the storage units provide the time-shifting capability. The results confirm the optimal economic operation of storage units with high efficiency. The use of a conventional power plant which has large CO_2 emission can be avoided by operating the storage units during peak-load hour.

3.6.8 Transmission Line Curtailment

In this application, energy storage units are integrated with the power system to supply energy in the time range of 6-12 hours. The wind generator needs to disconnect from the system due to several factors like a technical limitation in the transmission system or confirming the stability of the overall power system. In this case, storage units store energy and supply energy to the system in a regulated mode based on stability concerns and transmission system capability as a result disconnection of the generator can be avoided. Energy storage units like CAES, hydrogen storage unit, PHS system and flow batteries are suitable for this operation.

The transmission curtailment by using CAES and hydro-wind systems is briefly reported in [85, 86]. Generally, isolated wind conversion systems integrated with weak utility systems are the most challenging task for the researcher community. The use of storage units in a highly renewable penetration wind energy system minimizes wind curtailment, losses in the transmission lines, and number of storage units. It confirms eliminating the requirement of the new transmission line, reducing overall cost and security of supply.

The hydrogen-based storage technologies can also be used in wind turbine operation as rejected wind power produces a large amount of hydrogen. The hydrogen can be employed in various manners once it is stored. It can be either used in the fuel cell for electricity generation and supply to the electrical system during peak hour demand or other applications like the field of mobility. However, these storage units offer challenges like their economic feasibility [87].

3.6.9 Load Following

In this application, energy storage units are integrated with the power system to supply energy in the time range of a few minutes to 12 hours. There is mismatching between generation and load demand of wind energy system due to the chaotic properties of wind. This may cause numerous technical challenges like the stable operation of the electrical power systems. This phenomenon may also lead the frequency and voltage fluctuation in the network. In this scenario, storage units are employed for storing and injecting power into the electrical network for a long duration. The storage units like CAES, hydrogen storage unit, PHS system and flow batteries are suitable for this operation.

A combined wind generator and battery technology are implemented in Futumata, Japan. Here, a wind generator of 51 MW is combined with a NaS battery bank of 35 MW [88]. The bank is controlled in such a way that it not only overcomes the technical challenges but also addresses economic issues. An optimal energy management scheme is reported in [89] which focuses on the techno-economic benefit of the system. Here, the energy is stored during off-peak demand where the cost of the energy is very low and supplies the high-cost energy during peak demand hour.

Another storage unit called flow batteries is also commonly used for this application, the control and sizing aspect of which is briefly addressed in [88]. The techno-economic analysis for wind-flow batteries hybrid system is briefly discussed in this research work. The energy flow from the banks is controlled by using proper control techniques which enhances the forecasting of wind generator power; as a result the cost of storage requirement for grid connection can be reduced. Two storage battery banks of the capacity of 35 MW and 40 MW are used to improve the predictability of the output power of a wind generator of 100 MW capacity which is discussed in [89].

3.6.10 Unit Commitment

In this application, storage units are integrated with the power system to supply energy in the time range of hours to days. Due to the volatile nature of the wind, it is difficult to generate the rated power to meet the expected demand throughout the day. The forecast errors can be overcome by maintaining a proper level of energy reserves.

Hence, the integration of high-energy storage units with the network is suitable to overcome the forecasting errors and minimizes the use of storage units under normal operation. The high-capacity storage units like hydrogen-based technology, PHS system and CAES are suitable for this operation.

In this service, various optimization techniques are proposed to reduce the operational costs and increase the return of the installation [90, 91]. The unit commitment issues for CAES and wind turbine in the electrical power system are discussed in [91]. This approach can reduce the overall cost for the operation of the power system.

3.7 Conclusion

The HESS application in the wind energy system is briefly studied in this chapter. HESS plays a major role in islanded wind generators against various uncertainties and satisfies the balance in supply-demand side power management. Among all HESS configurations, battery-supercapacitors combination storage units give a tremendous performance as compared to other configurations due to their low cost, development of advanced technology and availability of different sizes in the market. More research may be carried out to verify the feasibility of different combination of storage units.

The various merits of HESS in wind energy applications are reported as a reduction in cost, more life period of the storage units, less dynamic response time, improved reliability, supply of momentary load, and enhanced power quality. The design of suitable topology may be required to search the benefits of an effective hybrid storage system.

Different HESS configurations are commonly used in wind energy applications. The cost and efficiency of the converters can be significantly improved by using impedance source converter topology. This HESS topology for wind farms has not been investigated yet, while the benefits of this topology are briefly reported in electric vehicles. However, the fewer number of switches in the converter, small volume, low cost, and more controllability can be achieved by using Z-source dc-ac power converters.

The coordination among the various storage units in HESS can be achieved by using an advanced control scheme like a hierarchical controller. Both centralized and decentralized control levels can be incorporated in hierarchical control methods, so that the delay time may affect the performance of the system which needs to be addressed.

Finally, the benefits of HESS application in wind farms are summarized by conducting a brief literature study. Numerous benefits are achieved by using HESS technology in wind power applications which is briefly summarized below:

- The power smoothing of wind generators can be obtained by integrating the storage units like flywheels, supercapacitors, or SMES. These storage systems have a high ramp rate and are helpful for reducing the frequency and voltage fluctuation in the electrical network.
- The optimal location of the short time scale storage units in the wind power plant and rating have been briefly discussed in this chapter. The topology of the converters used in the wind power plant and the control schemes used to regulate the power flow is also highlighted in this work.
- The stability of the overall system under disturbance such as LVDT facilities and oscillation damping issues can be enhanced by using HESS. In these applications, high ramp rate storage technologies are used to improve the grid code.
- The forecasting of the output power of the wind generator can be enhanced by using HESS. It also ensured the uninterrupted power supply for the critical load. The forecasting enhancement leads to both technical and economic benefits. It also reduces the operational expenses due to the diminished power reserve demand.

References

1. IRENA (2019), Future of wind: Deployment, investment, technology, grid integration and socio-economic aspects.
2. Council, G. W. E. (2016). Global wind report 2016–annual market update. Global Wind Energy Council: Brussels, Belgium.
3. Busby, R. L. (2012). *Wind power: The industry grows up*. PennWell Books.
4. Ayodele, T. R., & Ogunjuyigbe, A. S. O. (2015). Mitigation of wind power intermittency: Storage technology approach. *Renewable and Sustainable Energy Reviews*, 44, 447–456.
5. Rahman, M. M., Oni, A. O., Gemechu, E., & Kumar, A. (2020). Assessment of energy storage technologies: A review. *Energy Conversion and Management*, 223, 113295.
6. Ren, G., Liu, J., Wan, J., Guo, Y. and Yu, D., 2017. Overview of wind power intermittency: Impacts, measurements, and mitigation solutions. *Applied Energy*, 204, pp.47–65.
7. Farhadi, M., & Mohammed, O. (2015). Energy storage technologies for high-power applications. *IEEE Transactions on Industry Applications*, 52(3), 1953–1961.
8. Moghadasi, A. H., Heydari, H., & Farhadi, M. (2010). Pareto optimality for the design of SMES solenoid coils verified by magnetic field analysis. *IEEE Transactions on Applied Superconductivity*, 21(1), 1–20.
9. Akram, U., Nadarajah, M., Shah, R., & Milano, F. (2020). A review on rapid responsive energy storage technologies for frequency regulation in modern power systems. *Renewable and Sustainable Energy Reviews*, 120, 109626.
10. Abbey, C., & Joos, G. (2007). Supercapacitor energy storage for wind energy applications. *IEEE Transactions on Industry Applications*, 43(3), 769–776.
11. Zhu, H., Li, H., Liu, G., Ge, Y., Shi, J., Li, H., & Zhang, N. (2020). Energy storage in high renewable penetration power systems: Technologies, applications, supporting policies and suggestions. *CSEE Journal of Power and Energy Systems*.
12. Wicki, S., & Hansen, E. G. (2017). Clean energy storage technology in the making: An innovation systems perspective on flywheel energy storage. *Journal of Cleaner Production*, 162, 1118–1134.
13. Ran, L., Xiang, D., & Kirtley, J. L. (2011). Analysis of electromechanical interactions in a flywheel system with a doubly fed induction machine. *IEEE Transactions on Industry Applications*, 47(3), 1498–1506.
14. Yang, C. J. (2016). Pumped hydroelectric storage. In *Storing Energy* (pp. 25–38). Elsevier.
15. Rehman, S., Al-Hadhrami, L. M., & Alam, M. M. (2015). Pumped hydro energy storage system: A technological review. *Renewable and Sustainable Energy Reviews*, 44, 586–598.

16. Attya, A. B. T., & Hartkopf, T. (2015). Utilising stored wind energy by hydro-pumped storage to provide frequency support at high levels of wind energy penetration. *IET Generation, Transmission & Distribution*, 9(12), 1485–1497.
17. Singh, R., & Bansal, R. C. (2018). Review of HRESs based on storage options, system architecture and optimisation criteria and methodologies. *IET Renewable Power Generation*, 12(7), 747–760.
18. Abdellatif, D., AbdelHady, R., Ibrahim, A.M. et al. Conditions for economic competitiveness of pumped storage hydroelectric power plants in Egypt. Renewables 5, 2 (2018). https://doi.org/10.1186/s40807-018-0048-1 and is used under Creative Commons Attribution 4.0 International License (http://creativecommons.org/licenses/by/4.0/).
19. Wang, J., Lu, K., Ma, L., Wang, J., Dooner, M., Miao, S., ... & Wang, D. (2017). Overview of compressed air energy storage and technology development. *Energies*, 10(7), 991.
20. Boicea, V. A. (2014). Energy storage technologies: The past and the present. *Proceedings of the IEEE*, 102(11), 1777–1794.
21. Xing Luo, Jihong Wang, Mark Dooner, Jonathan Clarke, Overview of current development in electrical energy storage technologies and the application potential in power system operation, *Journal of Applied Energy*, Volume 137, pp. 511 – 536, 2015, ISSN:0306-2619,"https://doi.org/10.1016/j.apenergy.2014.09.081,(http://www.sciencedirect.com/science/article/pii/S0306261914010290) and is used under a CC-BY-3.0, (http://creativecommons.org/licenses/by/3.0/).
22. Zhang, S., Guo, W., Yang, F., Zheng, P., Qiao, R., & Li, Z. (2019). Recent Progress in Polysulfide Redox-Flow Batteries. *Batteries & Supercaps*, 2(7), 627–637.
23. Chen, H., Cong, T. N., Yang, W., Tan, C., Li, Y., & Ding, Y. (2009). Progress in electrical energy storage system: A critical review. *Progress in Natural Science*, 19(3), 291–312.
24. Wade, N.S., Taylor, P.C., Lang, P.D. and Jones, P.R., 2010. Evaluating the benefits of an electrical energy storage system in a future smart grid. *Energy Policy*, 38(11), pp. 7180–7188.
25. Hajiaghasi, S., Salemnia, A., & Hamzeh, M. (2019). Hybrid energy storage system for microgrids applications: A review. *Journal of Energy Storage*, 21, 543–570.
26. Ires, V. F., Romero-Cadaval, E., Vinnikov, D., Roasto, I., & Martins, J. F. (2014). Power converter interfaces for electrochemical energy storage systems–A review. *Energy Conversion and Management*, 86, 453–475.
27. Ju, F., Zhang, Q., Deng, W., & Li, J. (2014, August). Review of structures and control of battery-supercapacitor hybrid energy storage system for electric vehicles. In *2014 IEEE International Conference on Automation Science and Engineering (CASE)* (pp. 143–148). IEEE.

28. Jing, W., Lai, C.H., Wong, S.H.W. and Wong, M.L.D., 2016. Battery-supercapacitor hybrid energy storage system in standalone DC microgrids: areview. *IET Renewable Power Generation*, 11(4), pp. 461–469.
29. Zimmermann, T., Keil, P., Hofmann, M., Horsche, M. F., Pichlmaier, S., & Jossen, A. (2016). Review of system topologies for hybrid electrical energy storage systems. *Journal of Energy Storage*, 8, 78–90.
30. Lahyani, A., Venet, P., Guermazi, A., & Troudi, A. (2012). Battery/supercapacitors combination in uninterruptible power supply (UPS). *IEEE Transactions on Power Electronics*, 28(4), 1509–1522.
31. Zheng, J. P., Jow, T. R., & Ding, M. S. (2001). Hybrid power sources for pulsed current applications. *IEEE Transactions on Aerospace and Electronic Systems*, 37(1), 288–292.
32. Dougal, R. A., Liu, S., & White, R. E. (2002). Power and life extension of battery-ultracapacitor hybrids. *IEEE Transactions on Components and Packaging Technologies*, 25(1), 120–131.
33. Catherino, H. A., Burgel, J. F., Shi, P. L., Rusek, A., & Zou, X. (2006). Hybrid power supplies: A capacitor-assisted battery. *Journal of Power Sources*, 162(2), 965–70.
34. Song, Z., Hofmann, H., Li, J., Han, X., Zhang, X., & Ouyang, M. (2015). A comparison study of different semi-active hybrid energy storage system topologies for electric vehicles. *Journal of Power Sources*, 274, 400–411.
35. Cao, J., & Emadi, A. (2011). A new battery/ultracapacitor hybrid energy storage system for electric, hybrid, and plug-in hybrid electric vehicles. *IEEE Transactions on Power Electronics*, 27(1), 122–132.
36. Cohen, I. J., Wetz, D. A., Heinzel, J. M., & Dong, Q. (2014). Design and characterization of an actively controlled hybrid energy storage module for high-rate directed energy applications. *IEEE Transactions on Plasma Science*, 43(5), 1427–1433.
37. Momayyezan, M., Abeywardana, D. B. W., Hredzak, B., & Agelidis, V. G. (2016). Integrated reconfigurable configuration for battery/ultracapacitor hybrid energy storage systems. *IEEE Transactions on Energy Conversion*, 31(4), 1583–1590.
38. Mo, R., & Li, H. (2016). Hybrid energy storage system with active filter function for shipboard MVDC system applications based on isolated modular multilevel DC/DC converter. *IEEE Journal of Emerging and Selected Topics in Power Electronics*, 5(1), 79–87.
39. Bharadwaj, C. A., & Maiti, S. (2017, November). Modular multilevel converter based hybrid energy storage system. In *2017 IEEE PES Asia-Pacific Power and Energy Engineering Conference (APPEEC)* (pp. 1–6). IEEE.
40. Zhang, L., Tang, Y., Yang, S., & Gao, F. (2016, May). A modular multi-level converter-based grid-tied battery-supercapacitor hybrid energy storage system with decoupled power control. In *2016 IEEE 8th International Power Electronics and Motion Control Conference (IPEMC-ECCE Asia)* (pp. 2964–2971). IEEE.

41. Jiang, W., Xue, S., Zhang, L., Xu, W., Yu, K., Chen, W., & Zhang, L. (2017). Flexible power distribution control in an asymmetrical-cascaded-multi-level-converter-based hybrid energy storage system. *IEEE Transactions on Industrial Electronics*, 65(8), 6150–6159.
42. Etxeberria, A., Vechiu, I., Baudoin, S., Camblong, H., & Kreckelbergh, S. (2014). Control of a vanadium redox battery and supercapacitor using a three-level neutral point clamped converter. *Journal of Power Sources*, 248, 1170–1176.
43. Jayasinghe, S. G., Vilathgamuwa, D. M., & Madawala, U. K. (2011). Diode-clamped three-level inverter-based battery/supercapacitor direct integration scheme for renewable energy systems. *IEEE Transactions on Power Electronics*, 26(12), 3720–3729.
44. Piris-Botalla, L., Oggier, G. G., & García, G. O. (2017). Extending the power transfer capability of a three-port DC–DC converter for hybrid energy storage systems. *IET Power Electronics*, 10(13), 1687–1697.
45. Shreelekha, K., & Arulmozhi, S. (2016, March). Multiport isolated bidirectional DC-DC converter interfacing battery and supercapacitor for hybrid energy storage application. In *2016 International Conference on Electrical, Electronics, and Optimization Techniques (ICEEOT)* (pp. 2763–2768). IEEE.
46. Ding, Z., Yang, C., Zhang, Z., Wang, C., & Xie, S. (2013). A novel soft-switching multiport bidirectional DC–DC converter for hybrid energy storage system. *IEEE Transactions on Power Electronics*, 29(4), 1595–1609.
47. Khosrogorji, S., Ahmadian, M., Torkaman, H., & Soori, S. (2016). Multi-input DC/DC converters in connection with distributed generation units–A review. *Renewable and Sustainable Energy Reviews*, 66, 360–379.
48. Anno, T., & Koizumi, H. (2014). Double-input bidirectional DC/DC converter using cell-voltage equalizer with flyback transformer. *IEEE Transactions on Power Electronics*, 30(6), 2923–2934.
49. Tayab, U. B., Roslan, M. A. B., Hwai, L. J., & Kashif, M. (2017). A review of droop control techniques for microgrid. *Renewable and Sustainable Energy Reviews*, 76, 717–727.
50. Mirzaei, A., Jusoh, A., Salam, Z., Adib, E. and Farzanehfard, H., 2011. Analysis and design of a high efficiency bidirectional DC–DC converter for battery and ultracapacitor applications. *Simulation Modelling Practice and Theory*, 19(7), pp. 1651–1667.
51. Sharma, R. K., & Mishra, S. (2017). Dynamic power management and control of a PV PEM fuel-cell-based standalone ac/dc microgrid using hybrid energy storage. *IEEE Transactions on Industry Applications*, 54(1), 526–538.
52. Song, Q., & Chen, J. (2018, June). A decentralized energy management strategy for a battery/supercapacitor hybrid energy storage system in autonomous DC microgrid. In *2018 IEEE 27th International Symposium on Industrial Electronics (ISIE)* (pp. 19–24). IEEE.

53. Xu, Q., Xiao, J., Hu, X., Wang, P., & Lee, M. Y. (2017). A decentralized power management strategy for hybrid energy storage system with autonomous bus voltage restoration and state-of-charge recovery. *IEEE Transactions on Industrial Electronics*, 64(9), 7098–7108.
54. Zhang, Y., & Li, Y. W. (2016). Energy management strategy for supercapacitor in droop-controlled DC microgrid using virtual impedance. *IEEE Transactions on Power Electronics*, 32(4), 2704–2716.
55. Xu, Q., Hu, X., Wang, P., Xiao, J., Tu, P., Wen, C., & Lee, M. Y. (2016). A decentralized dynamic power sharing strategy for hybrid energy storage system in autonomous DC microgrid. *IEEE Transactions on Industrial Electronics*, 64(7), 5930–5941.
56. Shi, M., Chen, X., Zhou, J., Chen, Y., Wen, J., & He, H. (2018). Advanced secondary voltage recovery control for multiple HESSs in a droop-controlled DC microgrid. *IEEE Transactions on Smart Grid*, 10(4), 3828–3839.
57. Bocklisch, T. (2016). Hybrid energy storage approach for renewable energy applications. *Journal of Energy Storage*, 8, 311–319.
58. Wang, B., Manandhar, U., Zhang, X., Gooi, H. B., & Ukil, A. (2018). Deadbeat control for hybrid energy storage systems in DC microgrids. *IEEE Transactions on Sustainable Energy*, 10(4), 1867–1877.
59. Jin, Z., Meng, L., Guerrero, J. M., & Han, R. (2017). Hierarchical control design for a shipboard power system with DC distribution and energy storage aboard future more-electric ships. *IEEE Transactions on Industrial Informatics*, 14(2), 703–714.
60. Cohen, I. J., Wetz, D. A., McRee, B. J., Dong, Q., & Heinzel, J. M. (2017). Fuzzy logic control of a hybrid energy storage module for use as a high rate prime power supply. *IEEE Transactions on Dielectrics and Electrical Insulation*, 24(6), 3887–3893.
61. Shi, J., Huang, W., Tai, N., Qiu, P., & Lu, Y. (2017). Energy management strategy for microgrids including heat pump air-conditioning and hybrid energy storage systems. *Journal of Engineering*, 2017(13), 2412–2416.
62. Yin, H., Zhou, W., Li, M., Ma, C., & Zhao, C. (2016). An adaptive fuzzy logic-based energy management strategy on battery/ultracapacitor hybrid electric vehicles. *IEEE Transactions on Transportation Electrification*, 2(3), 300–311.
63. Hredzak, B., Agelidis, V. G., & Jang, M. (2013). A model predictive control system for a hybrid battery-ultracapacitor power source. *IEEE Transactions on Power Electronics*, 29(3), 1469–1479.
64. Chettibi, N., Mellit, A., Sulligoi, G., & Pavan, A. M. (2016). Adaptive neural network-based control of a hybrid AC/DC microgrid. *IEEE Transactions on Smart Grid*, 9(3), 1667–1679.
65. Barton, J. P., & Infield, D. G. (2004). Energy storage and its use with intermittent renewable energy. *IEEE Transactions on Energy Conversion*, 19(2), 441–448.

66. Bayod-Rújula, A. A. (2009). Future development of the electricity systems with distributed generation. *Energy*, 34(3), 377–383.
67. Muljadi, E., Butterfield, C. P., Chacon, J., & Romanowitz, H. (2006, June). Power quality aspects in a wind power plant. In *2006 IEEE Power Engineering Society General Meeting* (pp. 8-pp). IEEE.
68. Qu, L., & Qiao, W. (2010). Constant power control of DFIG wind turbines with supercapacitor energy storage. *IEEE Transactions on Industry Applications*, 47(1), 359–367.
69. Boukettaya, G., Krichen, L., & Ouali, A. (2010). A comparative study of three different sensorless vector control strategies for a Flywheel Energy Storage System. *Energy*, 35(1), 132–139.
70. Cardenas, R., Pena, R., Asher, G., & Clare, J. (2001). Control strategies for enhanced power smoothing in wind energy systems using a flywheel driven by a vector-controlled induction machine. *IEEE Transactions on Industrial Electronics*, 48(3), 625–635.
71. Cárdenas, R., Peña, R., Pérez, M., Clare, J., Asher, G., & Wheeler, P. (2006). Power smoothing using a flywheel driven by a switched reluctance machine. *IEEE Transactions on Industrial Electronics*, 53(4), 1086–1093.
72. Ray, P. K., Mohanty, S. R., & Kishor, N. (2011). Proportional–integral controller based small-signal analysis of hybrid distributed generation systems. *Energy Conversion and Management*, 52(4), 1943–1954.
73. Gomis-Bellmunt, O., Liang, J., Ekanayake, J., & Jenkins, N. (2011). Voltage–current characteristics of multiterminal HVDC-VSC for offshore wind farms. *Electric Power Systems Research*, 81(2), 440–450.
74. Baroudi, J. A., Dinavahi, V., & Knight, A. M. (2007). A review of power converter topologies for wind generators. *Renewable energy*, 32(14), 2369–2385.
75. Abbey, C., & Joos, G. (2007). Supercapacitor energy storage for wind energy applications. *IEEE Transactions on Industry Applications*, 43(3), 769–776.
76. Barton, J. P., & Infield, D. G. (2004). Energy storage and its use with intermittent renewable energy. *IEEE Transactions on Energy Conversion*, 19(2), 441–448.
77. Brekken, T. K., Yokochi, A., Von Jouanne, A., Yen, Z. Z., Hapke, H. M., & Halamay, D. A. (2010). Optimal energy storage sizing and control for wind power applications. *IEEE Transactions on Sustainable Energy*, 2(1), 69–77.
78. Roberts, B. P. (2008, July). Sodium-Sulfur (NaS) batteries for utility energy storage applications. In *2008 IEEE Power and Energy Society General Meeting-Conversion and Delivery of Electrical Energy in the 21st Century* (pp. 1–2). IEEE.
79. Dufo-López, R., Bernal-Agustín, J. L., & Domínguez-Navarro, J. A. (2009). Generation management using batteries in wind farms: Economical and technical analysis for Spain. *Energy Policy*, 37(1), 126–139.
80. Bernal-Agustín, J. L., & Dufo-López, R. (2008). Hourly energy management for grid-connected wind–hydrogen systems. *International Journal of Hydrogen Energy*, 33(22), 6401–6413.

81. Papaefthymiou, S. V., Karamanou, E. G., Papathanassiou, S. A., & Papadopoulos, M. P. (2010). A wind-hydro-pumped storage station leading to high RES penetration in the autonomous island system of Ikaria. *IEEE Transactions on Sustainable Energy*, 1(3), 163–172.
82. Mercier, P., Cherkaoui, R., & Oudalov, A. (2009). Optimizing a battery energy storage system for frequency control application in an isolated power system. *IEEE Transactions on Power Systems*, 24(3), 1469–1477.
83. Sasaki, T., Kadoya, T., & Enomoto, K. (2004). Study on load frequency control using redox flow batteries. *IEEE Transactions on Power Systems*, 19(1), 660–667.
84. Nyamdash, B., Denny, E., & O'Malley, M. (2010). The viability of balancing wind generation with large scale energy storage. *Energy Policy*, 38(11), 7200–7208.
85. Anagnostopoulos, J.S. and Papantonis, D.E., (2008). Simulation and size optimization of a pumped–storage power plant for the recovery of windfarms rejected energy. *Renewable Energy*, 33(7), pp. 1685–1694.
86. Denholm, P., & Sioshansi, R. (2009). The value of compressed air energy storage with wind in transmission-constrained electric power systems. *Energy Policy*, 37(8), 3149–3158.
87. Loisel, R., Mercier, A., Gatzen, C., Elms, N., & Petric, H. (2010). Valuation framework for large scale electricity storage in a case with wind curtailment. *Energy Policy*, 38(11), 7323–7337.
88. Iijima, Y., Sakanaka, Y., Kawakami, N., Fukuhara, M., Ogawa, K., Bando, M., & Matsuda, T. (2010, June). Development and field experiences of NAS battery inverter for power stabilization of a 51 MW wind farm. In *The 2010 International Power Electronics Conference-ECCE ASIA* (pp. 1837–1841). IEEE.
89. Hida, Y., Yokoyama, R., Shimizukawa, J., Iba, K., Tanaka, K., & Seki, T. (2010, July). Load following operation of NAS battery by setting statistic margins to avoid risks. In *IEEE PES general meeting* (pp. 1–5). IEEE.
90. Brown, P. D., Lopes, J. P., & Matos, M. A. (2008). Optimization of pumped storage capacity in an isolated power system with large renewable penetration. *IEEE Transactions on Power Systems*, 23(2), 523–531.
91. Daneshi, H., Daneshi, A., Tabari, N. M., & Jahromi, A. N. (2009, May). Security-constrained unit commitment in a system with wind generation and compressed air energy storage. In *2009 6th International Conference on the European Energy Market* (pp. 1–6). IEEE.

4

Advances in Electrochemical Energy Storage Device: Supercapacitor

Swagatika Kamila[1,2], Bikash Kumar Jena[1,2*] and Suddhasatwa Basu[1,2†]

[1]*Materials Chemistry Department, CSIR-Institute of Minerals and Materials Technology, Bhubaneswar, India*
[2]*Academy of Scientific & Innovative Research (AcSIR), Ghaziabad, India*

Abstract

The storage of electrical energy is an essential technology in recent years. It builds an application with future renewable energy-based technology, hybrid electric vehicles, and the manufacturing of portable electronic devices. The supercapacitor is an important energy storage device due to its rapid charge-discharge process, longer cycle life (>100000 cycles), and high power density compared to rechargeable batteries and more energy density than the conventional capacitors. A supercapacitor stores electrical charge by electrostatic ion adsorption process and pseudocapacitance process involving fast surface redox reaction. However, getting the energy density performance of a supercapacitor close to the rechargeable batteries is still challenging. Better energy storage performance can be achieved by developing new nanostructure electrode materials (having both capacitive type and battery type property), using stable electrolytes with high cell potential and design of hybrid supercapacitor device. This chapter provides an overview of the basic concept and principle of electrochemical energy storage devices (supercapacitor) along with the advancement of the materials developed, properties, and applications.

Keywords: Energy storage device, supercapacitor, energy density, power density, hybrid supercapacitor device

---

*Corresponding author: bikash@immt.res.in
†Corresponding author: sbasu@immt.res.in

4.1 Introduction

Energy storage plays a key role in storing the available resources of energy and utilizing them for future use. Energy storage is really necessary if the primary source of energy is only available on an intermittent basis like renewable sources. Energy exists in different forms in nature like electrical, thermal, chemical, nuclear, radiant, gravitational energy, etc., but the most important form is electrical energy. The real necessity for energy storage is to generate large-scale electricity to meet the energy demands in modern society. Electricity is the flow of energy which travels very fast, and it has no shelf-life. Once it is generated from an energy reservoir, it is instantaneously used by energy-consuming devices and the rest is lost. So its storage is important for the future use whenever needed.[1-3] Now research has been devoted towards the development of clean energy storage technology instead of the consumption of fossil fuel which causes a detrimental effect on society. The storage technology is important for the future of renewable energy–based technology, hybrid electric vehicles and manufacturing of portable electronic devices. The electricity generated from solar and wind energy fluctuates dramatically. It is difficult to couple the electricity provider with the grid. However, grids require stable power. Here, the importance of energy storage devices that function by installing batteries into the electrical grids, so that it can be added easily with renewable energy.[4,5] The batteries store the renewable energy and deliver it to the grids whenever needed. This makes the power supply smoother and more predictable. Further, the areas not having the facility of grid electricity supply, their energy storage devices can be utilized to get electricity from renewable sources. In transportation, energy storage devices have been used for the development of hybrid energy vehicles (HEVs). In HEVs the electrical storage device has been hybrid with the small petrol engine, so that there is a reduction of fossil fuel consumption. Some HEVs use regenerative braking, in which the electric motor works as a generator. While braking, some of the wheels' kinetic energy is transformed into electrical energy and stored in the battery. This energy is available for accelerating the vehicle again. In modern society, advanced electronic devices like smartphones, iPads, iPods, cameras, laptops, etc., are accessed by the help of lithium-ion batteries.

4.2 Types of Energy Storage Devices

The energy storage system accepts and dissipates the electrical energy, but it stores the energy by converting it into another form. The electrical energy storage devices are in various forms, and these are developed based on their energy storage principle. The energy storage technology includes various processes i.e, mechanical, thermal, chemical, electrical and electrochemical.[6] The various forms of electrical energy storage systems are presented in Figure 4.1.

Mechanical Energy Storage Device: It stores excess electrical energy in the form of kinetic energy in flywheels, potential energy in water and compression energy in the air and again converts it into electrical energy.

Thermal Energy Storage Device: Steam boiler is an example of a thermal energy storage device that produces energy by using coal, natural gas, biomass as fuels. The produced steam operates the turbine that is connected to a generator and electricity is generated.

Chemical Energy Storage Device: It stores energy in the form of chemical fuels. Hydrogen is a form of a fuel, produced electricity in fuel cell.

Electrical Energy Storage Device: These devices store electrical energy electrostatically. Capacitor and superconducting magnetic storage are examples of electrical energy storage device. Superconducting Magnetic Energy Storage (SMES) is the new technology which stores electricity from the grid within the magnetic field. A capacitor is a device which stores electrical charge. It is constructed by two conducting plates separated by a dielectric material. When a capacitor is attached to a power source, a potential difference arises between the two plates. So that an electric field

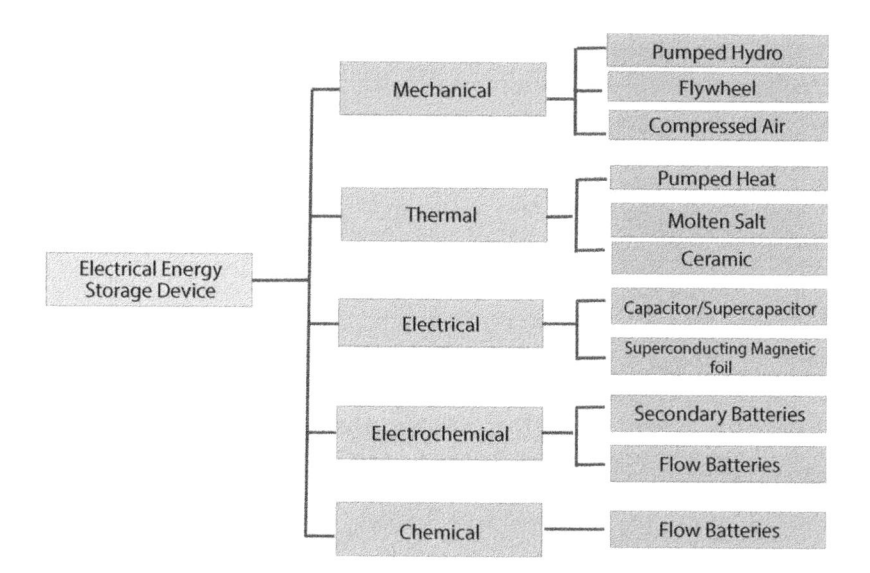

Figure 4.1 Different types of electrical energy storage systems.

develops across the dielectric material and the charge separation takes place at the electrode surface. Capacitor stores energy in the electric field exists between the two conducting plates.[7]

Electrochemical Energy Storage Device: The common feature of this device is primarily to store the chemical energy through the electrochemical oxidation-reduction reverse process, and then this chemical energy is converted into electrical energy. Batteries including secondary batteries, flow batteries and supercapacitors are the electrochemical energy storage devices. Some common commercially available secondary batteries are divided into the following groups: standard batteries including lead-acid and Ni-Cd, advanced batteries such as Li-ion, Li-ion polymer, Ni-metal hydride and special batteries including Ag-Zn and $Ni-H_2$. The flow batteries like Br_2-Zn and Vanadium redox batteries stores energy directly in the electrolyte solution.[8,9]

4.3 Overview of Supercapacitor and Its Global Scenario

Considering the reliable, stable and sustainable technologies, the supercapacitor has been recognized as an important class of electrical energy storage system. Supercapacitors are also known as an electrochemical capacitor or electrical double-layer capacitor. The design of a supercapacitor is similar to a battery that consists of two current collectors, electrode materials, electrolytes and an ion permeable membrane (Figure 4.2). But the main component of a supercapacitor device is the nanostructured electrode materials. A supercapacitor is important due to its rapid charge-discharge process, longer cycle life (>100000 cycles) and high power density

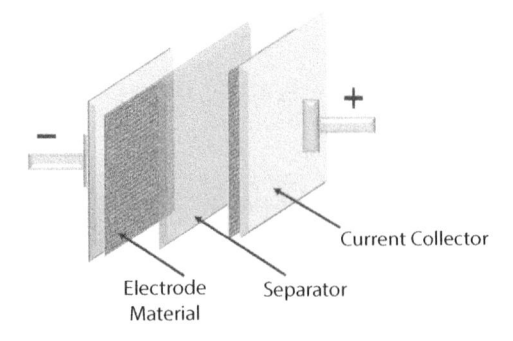

Figure 4.2 Schematic representation of supercapacitor.

compared to rechargeable batteries and higher energy density than the conventional capacitors.

It provides stabilized power supply to the fluctuating loads such as laptops, GPS and portable media players. Such advantages of the supercapacitor device make it useful in market demands. This revolutionary energy storage technology meets demands in hybrid energy vehicles, renewable energy storage devices, smart grids, emergency backup power to low power equipments such as RAM, SRAM, microcontroller and PC cards, etc., which establishes supercapacitor adoption globally.[10] The idea of the electrochemical capacitor has been known for many years. In 1957, Becker (General Electric) patented a low-voltage electrolytic capacitor using porous carbon electrode in an aqueous medium. In 1970, the SOHIO company patented another version of the electrolytic capacitor based on carbon electrode in the organic electrolyte. This patent was transformed into Nippon Electric Company (NEC), which first commercialized supercapacitor technology as memory backup for electronics. After that, there was enormous intensification of research on supercapacitor. Many companies have initiated the production of supercapacitors into the market. In 1982, Pinnacle Research Institute (PRI) developed a low internal resistance, portable and high power supercapacitor device. In 1992, Maxwell technology took over the PRI technology and developed a supercapacitor device named as boost cap. This continuous research on energy storage technology has resulted in today's high-performance supercapacitor device available commercially.[11] In today's market supercapacitors are available in various forms according to the device construction and shape. These vary from small chip and coin cell types to a large model like a cylindrical and rectangular shape (Figure 4.3).

The electrochemical performance of these devices also differs by its design and is used for a specific purpose. Table 4.1 represents the performance and application of various supercapacitor devices.[12]

The rapid technological advancements and evolving green energy application produce global demand for the supercapacitor. According to a research report, it is expected that the global supercapacitor market will reach $3.1 billion by 2026 and it is forecast to grow at a compound annual growth rate (CAGR) of 15.5% for the period 2017-26.[13] The global supercapacitor market is segmented according to technology, types, and products and by region.[14] On the basis of technology, supercapacitors are available based on organic electrolyte and aqueous electrolyte. But the organic electrolyte-based supercapacitors are more popular compared to others. There are mainly three types of supercapacitors: EDLC, pseudocapacitor and hybrid supercapacitor. The EDLC-based supercapacitors are dominating

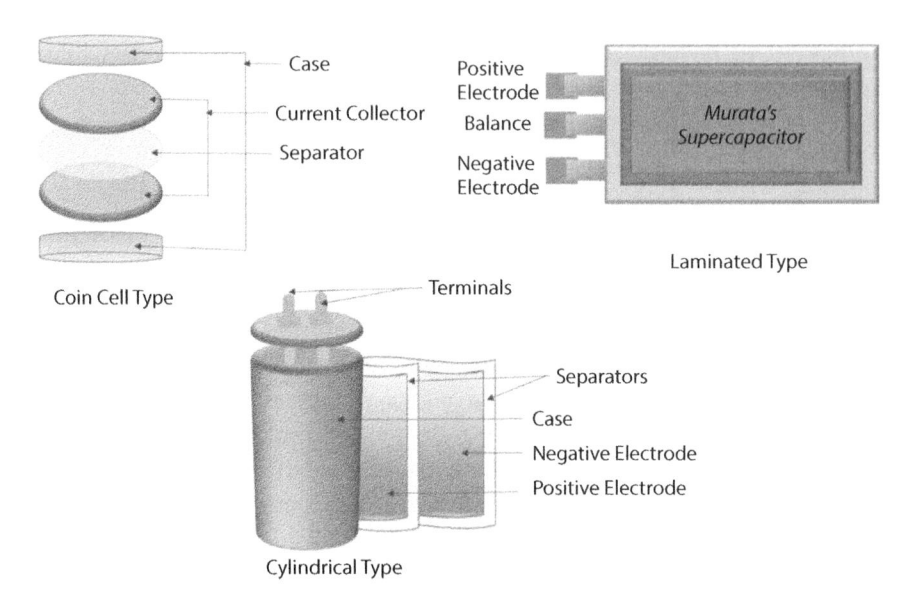

Figure 4.3 Schematic representation of various designs of supercapacitor devices.

the market and their applications vary from electronics to the transport sector. However, it is expected that demand for the hybrid supercapacitor will experience a higher growth rate in the future due to its ability to maintain both high energy and power density performance. With the consideration of its applications, the global supercapacitor market is segmented into the consumer electronics, transportation, industrial automation, medical, energy and power sectors. The high demand for electronic gadgets increases the supercapacitor market compared to other fields. The global supercapacitor market is segmented into different regions such as North America, Europe, Asia Pacific and others. In 2015, North America held the top rank in the global supercapacitor market with 39% share due to high demands in electronic gadgets. However, this market has reached saturation and the growth of the supercapacitor market is getting sluggish in North America during the forecast period. In Europe, the supercapacitor market has been promoted owing to the rapid development of the automobile industry in some major countries. Now Asia Pacific is the fastest-growing supercapacitor market due to the increasing demands in transport, consumer electronics, and rapid industrialization especially in India and China. Some of the companies playing in this supercapacitor market are AVX Corp., CAP-XX, Elna Co. Ltd, Graphene Laboratories INC, Axiom Power International, Elton, Maxwell Technologies INC, Panasonic Co. Ltd., and Skeleton Technologies.

Table 4.1 Types of double-layer capacitors.*

SC device	Specific capacitance	Application
Chip type	Ultra-low capacitance 0.1F or less	Used in memory backup
Coin cell type		
Punch type	Ultra-low capacitance 0.1F to 1F	Standby power for home appliances
Laminated type		Load levelling of batteries in mobile devices to extend the durability, High-current for driving high-brightness flash LEDs, Production of energy
Cylindrical type	Medium capacitance 1F to 100F	Road marking studs, LED signs, drive power in compact motor in toys
Rectangular type		
Modules (interconnected cylindrical and rectangular cells)	Capacitance higher than 100F	Energy regeneration in industrial equipments and automobiles, emergency power supply in renewable energy generation control

*(https://www.global.tdk.com/techmag/electronics_primer/vol8.htm).

4.4 Status of Supercapacitor in India

India is the fastest-growing economy in the world today. India's supercapacitor market is forecast at a CAGR of 16% in the period of 2017-22 due to the huge demands in various sectors.[15] India has developed supercapacitor having different voltage by using naturally occurring material and specifically used it in the defense and space sectors. Centre for materials and electronics and information technology under the Department of Electronics and Technology (DET) developed the supercapacitors and sent them for trials to state-run defense and space agencies in the country. After it is standardized for strategic use, this technology was transferred to other private companies for large-scale production. India is the world's fourth-largest market for the automobile sector.

India Supercapacitors Market Size, By End User, By Value, 2012-2022F (USD Million)

TechSci Research
from NOW to NEXT

2012 2013 2014 2015 2016 2017E 2018F 2019F 2020F 2021F 2022F

■ Consumer Electronics   ■ Energy   ■ Industrial   ■ Automotive   ■ Others

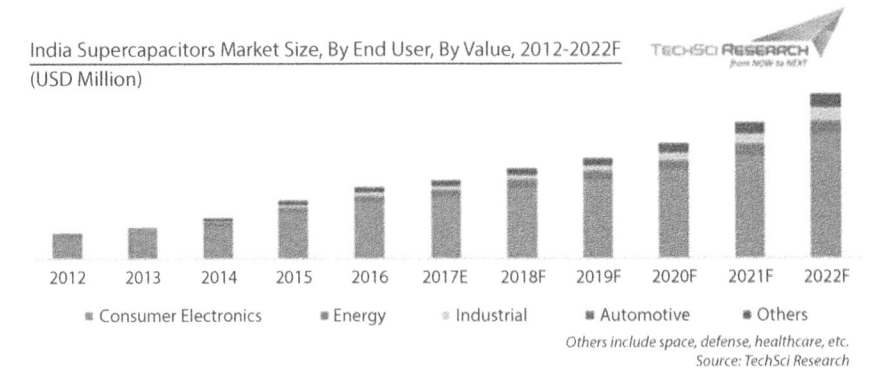

Others include space, defense, healthcare, etc.
Source: TechSci Research

Figure 4.4 The TechSci research provided a clear database of India's supercapacitor market size from 2012-2022 periods. (https://www.techsciresearch.com/report/india-supercapacitors-market/1370.html).

The government of India launched the National Electric and Mobility Plan (NEMMP), with an aim to end the fuel crisis and reduce environmental pollution by promoting eco-friendly vehicles in India. It is assumed that about 5 to 7 million electric vehicles have been launched by 2020 in India. Tata Motors implemented the supercapacitor technology for hybrid electric vehicles and collaborated with other private firms such as Spell and Chheda Electrical at Pune and Aartech Solonics at Bhopal in Madhya Pradesh to increase production and localize the global technology. The TechSci research represents a clear picture of India's supercapacitor market size from 2012-2022 (Figure 4.4).

4.5 Types of Supercapacitor According to the Energy Storage Mechanism

In supercapacitor device charge separation occurs from the electrolyte to the surface of electrode materials and it stores charge through different processes. By considering the energy storage processes, supercapacitors are divided into three types: electrical double-layer capacitor (EDLC), pseudo-capacitor and hybrid supercapacitor (Figure 4.5).[16]

4.5.1 Electrical Double-Layer Capacitor (EDLC)

The energy storage principle of EDLC involves the electrostatic reversible ion adsorption process in electrolyte onto active materials. The charge

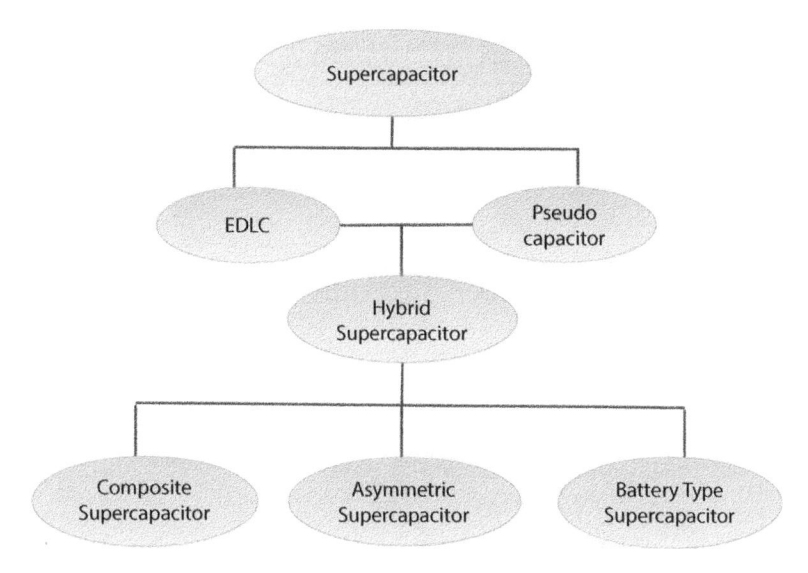

Figure 4.5 Diagram of different types of supercapacitor.

separation occurs under the polarization of electrode. Under this condition, the oppositely charged ions diffuse through the electrolyte and produce a condensed layer having a thickness of a few nanometres at the electrode surface to maintain the charge neutrality. This layer is known as the electrical double layer (EDLC) explained by Helmholtz and it acts as a dielectric medium. This Helmholtz layer is simply called an electrical capacitor and is presented in Figure 4.6a. However, the Helmholtz model fails to explain the diffusion process and the interaction of dipole moment of the solvent molecule with the electrode. So Gouy and Chapman proposed a diffusion model of EDL with the concept that the ions are mobile in electrolytes due to the combined effect of diffusion and electrostatic force. This is known as diffuse layer (Figure 4.6b). However, the Gouy and Chapman model is insufficient for highly charged EDL. In 1924, a new EDL model was proposed by Stern that fused both the Helmholtz and Gouy-Chapman model and described the behavior of EDL capacitor with using two regions i.e, inner and outer region. The inner region of thickness H is termed as the stern layer and the outer one is termed as diffuse layer and presented in Figure 4.6c.[17,18] The reversible electrostatic process gives the charging and discharging states of EDLC. The EDLC consist of electrodes, electrolytes, and a separator which prevents it from the electronic contact of cathode and anode inside the electrolyte. By considering the design and principle of EDLC, an equivalent circuit diagram is shown in Figure 4.7. Generally conducting and high surface area carbon-based materials are

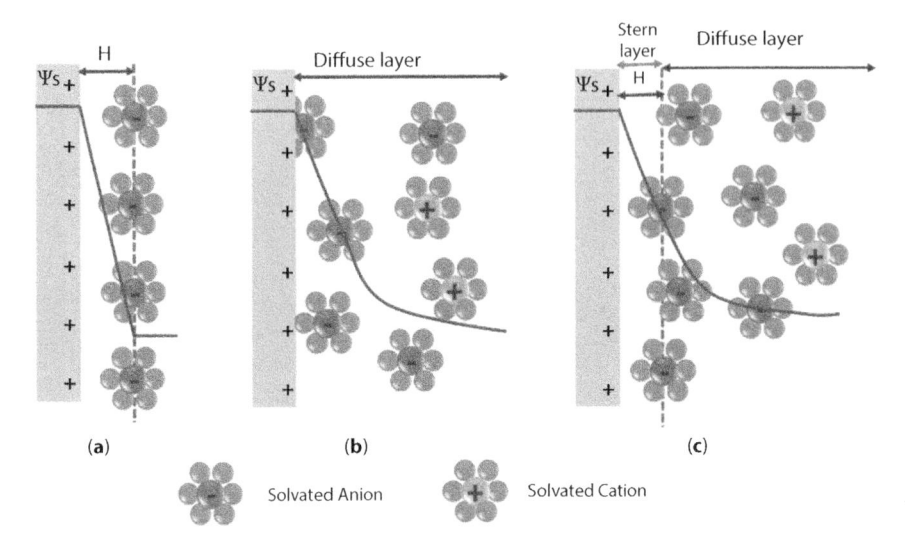

Figure 4.6 Schematic representation of (a) Helmholtz model, (b) Gouy-Chapman model and (c) Gouy-Chapman-Stern model of EDLC supercapacitor.

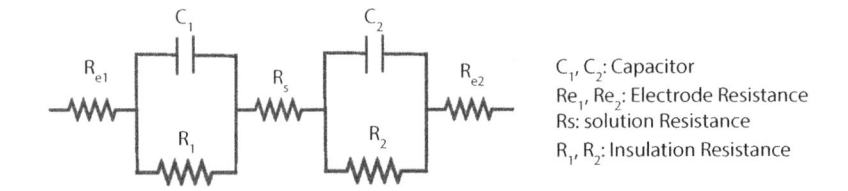

Figure 4.7 Equivalent circuit diagram of EDLC supercapacitor.

used in EDLC devices. The high surface area based porous carbon material is generally employed in an EDLC supercapacitor device.

4.5.2 Pseudocapacitor

A pseudocapacitor stores electrical energy by electron transfer process between the electrode material and electrolyte. This faradic process involved in supercapacitor gives better capacitance and energy density performance compared to EDLC supercapacitor. Pseudocapacitance arises due to thermodynamic reason and is potential dependent. There are three types of chemical processes which occur to develop a pseudocapacitor, i.e., surface adsorption or chemisorption of ions from the electrolyte, faradic reaction of transition metal oxides and reversible electrochemical doping-dedoping

ELECTROCHEMICAL ENERGY STORAGE DEVICE: SUPERCAPACITOR 129

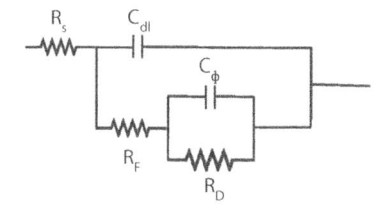

Figure 4.8 Equivalent circuit diagram of pseudocapacitor.

process in conducting polymer-based electrodes.[19] Generally transition metal oxides/hydroxides, transition metal sulfides and conducting polymers are used as pseudocapacitive electrode material in supercapacitor application.

By considering the design and principle of pseudocapacitance, an equivalent electrical circuit diagram is presented in Figure 4.8. Where C_Φ is the pseudocapacitance, R_F is the electrode and electrolyte resistance and R_D is the faradic resistance during discharge. Usually, over a certain range of potential, C_Φ may well exceed the double layer capacitance. The parallel combination of faradic impedance (C_Φ, R_F and R_D) with C_{dl} helps to build up capacitance of pseudocapacitor because of the additive law of capacitance.

4.5.3 Hybrid Supercapacitor

As we know, EDLC electrode proves good cycle life and high power delivery, but the pseudocapacitor gives high energy density. So hybrid supercapacitor is the combination of both pseudocapacitor as high energy source and the EDL capacitor as a power source. The main purpose for the design of hybrid supercapacitor is to increase both energy and power rate performance in a wide cell potential. There are three types of hybrid supercapacitor distinguished by the electrode configuration: composite supercapacitor, asymmetric supercapacitor and battery type supercapacitor.[20]

4.5.3.1 *Composite Supercapacitor*

Composite supercapacitors are made up both carbon material combines with the pseudocapacitive electrode material like metal oxide/hydroxides and conducting polymers. This combined electrode material is used in single electrode. The charge storage involves both electrostatic and chemical process.[21,22]

4.5.3.2 Asymmetric Supercapacitor

An asymmetric supercapacitor involves both faradic and non-faradic processes by coupling EDLC and pseudocapacitive electrode in a cell. Generally, carbon-based materials are taken as cathode and the pseudocapacitive materials are taken as anode. A prototype of supercapacitor was invented by Ammatucci *et al.* in 2001 and later further research was conducted by other groups.[23,24]

4.5.3.3 Battery Type

A battery type device combines two different electrodes, such as the battery type electrode and supercapacitor electrode assembled into one device. In this design, both properties of battery and supercapacitor is implemented.[25]

4.6 Basic Components of Supercapacitor

4.6.1 Current Collector

The choice of the current collector is an important factor for the fabrication of supercapacitor device. The chemical stability of the current collector in specific electrolytes and various morphology is effective for the long cycle life and better charge storage performance of the device. In the case of strong acidic electrolyte, some anti-corrosion materials like Au, ITO (indium tin oxide), carbon-based current collectors have been used.[26-28] For the alkaline electrolyte, Ni, Ti, carbon and stainless steel materials are used. Among these, Ni-based current collectors are commonly used in alkaline electrolyte because of its low cost and good stability. It gives additional pseudocapacitance behavior due to the generation of Ni oxide/hydroxide on it. Ni foam is very good current collector, due to its mesh-like morphology with more surface area, so that the active material is well exposed towards the electrolytes for better capacitance properties.[29] In neutral electrolyte, Ni, Ti, stainless steel, carbon-based current collectors such as carbon cloth, carbon fibre paper, ultrathin-graphite foams are used.[30-32] The less corrosive nature of the neutral electrolyte is more favorable for the supercapacitor current collector material. In the case of the organic electrolyte, most of Al is used as current collector.[33]

4.6.2 Electrode Materials

The capacitance of a supercapacitor device is intimately dependent upon the electrode materials used. So, it is important to develop high surface area nanostructured electrode materials for supercapacitor application. The material should be conducting/semiconducting in nature. Also the electrochemical stability of the electrode material is important for long cycle life of the device. There are two types of electrode materials developed for supercapacitor application such as EDLC electrode material and pseudocapacitive (faradic) electrode material.

4.6.2.1 EDLC Materials

Nanostructured carbon-based materials are widely accepted for EDL supercapacitor. The low cost, conducting in nature, large surface area, chemically inert and naturally abundance make these carbon-based supercapacitor for commercialization.[34] The different types of EDLC electrode materials are presented in Figure 4.9. The carbon-based materials also work in a wide temperature range and are electrochemically stable at a wide potential range, which makes the supercapacitor a high voltage device. Normally, carbon-based materials store charge at the electrical double layer. So capacitance properties dependent of the large surface area

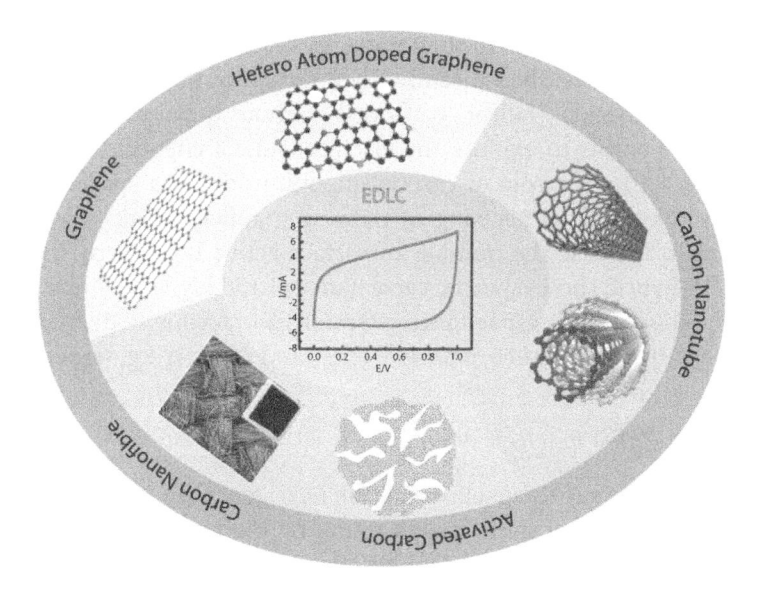

Figure 4.9 Schematic representation of different types of EDLC electrode materials.

based electrode materials. The surface area of carbon-based materials can be increased by doing some modification such as chemical activation of carbon by strong alkali, synthesis of porous carbon nanostructures, functionalization of carbon surface and the doping of hetero atoms (N, S, B, P). The design of different morphology of carbon material like carbon nanotube (CNT), carbon nanofibre (CNF), carbon nanosphere are also used for supercapacitor application.[35] A wide range of high surface area carbon materials are investigated for the energy storage application. The capacitance of activated carbon typically ranges from 40-140 F/g and 15-135 F/g for carbon nanotube.[36-38] Currently the best capacitance result of 130 F/g is developed by the commercial supercapacitor developer, Maxwell boost cap. The discovery of graphene from graphite in 2004, has opened a new era for the study of 2D layered material in the energy storage field. The layered like structure, high surface area, high conductivity, chemical inertness makes it active in the field of energy storage. In principle, graphene with its theoretical surface area, 2675 m^2/g gives specific capacitance of 550 F/g. However, it is not practically possible to get such value of the surface area. The average value of reduced graphene shows in the range 300-1000 m^2/g which gives the lower value of capacitance in the range of 100-270 F/g and 70-120 F/g at aqueous and organic electrolyte, respectively.[39,40] However, the restacking property of graphene decreases the surface area as well as the capacitance property. So to avoid this detrimental effect some chemical activation, exfoliation, development of vertically aligned graphene and making of the composite with CNT/carbon nanosphere/carbon quantum dots have been carried out for better energy storage performance.[41-43] Jena and co-workers developed a sandwiched structure of reduced graphene oxide and Nitrogen, Sulfur co-doped carbon quantum dots (rGO/CQDs-HP) hybrids to explore the electrochemical energy storage activity (Figure 4.10). The role of carbon quantum dots as a spacer material that prevents the layer stacking of reduced graphene oxides nanosheets and helps to increase the number of surface active sites. Here, the rGO/CQDs-HP hybrid shows specific capacitance of 156 F/g with robust operational stability of (90% capacitance retention) and having high energy and power density performance of 22.4 W h kg^{-1} and 76.68 W kg, respectively.[44]

4.6.2.2 *Pseudocapacitive Materials*

Pseudocapacitive supercapacitors are the next-generation supercapacitors due to its high value of specific capacitance i.e, higher than 1000 F/g. They not only store charge in the EDL, but also the reversible faradic process occurs at the bulk of the electrode material which gives a high value of

ELECTROCHEMICAL ENERGY STORAGE DEVICE: SUPERCAPACITOR 133

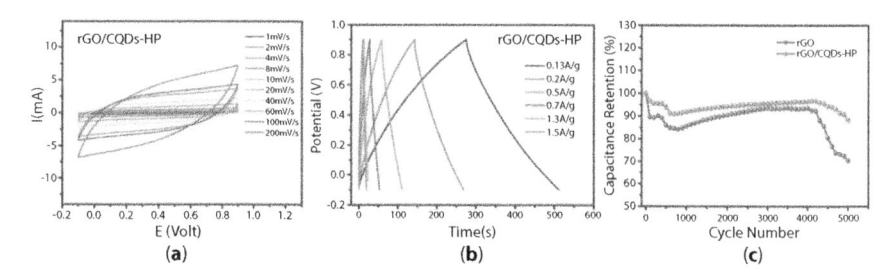

Figure 4.10 (a) Overlap CV plot of rGO/CQDs-HP at different scan rates (b) overlap CD plot of rGO/CQDs-HP at different current densities and (c) operational stability of rGO/CQDs-HP over 5000 cycles [Reproduced with permission from ref 44].

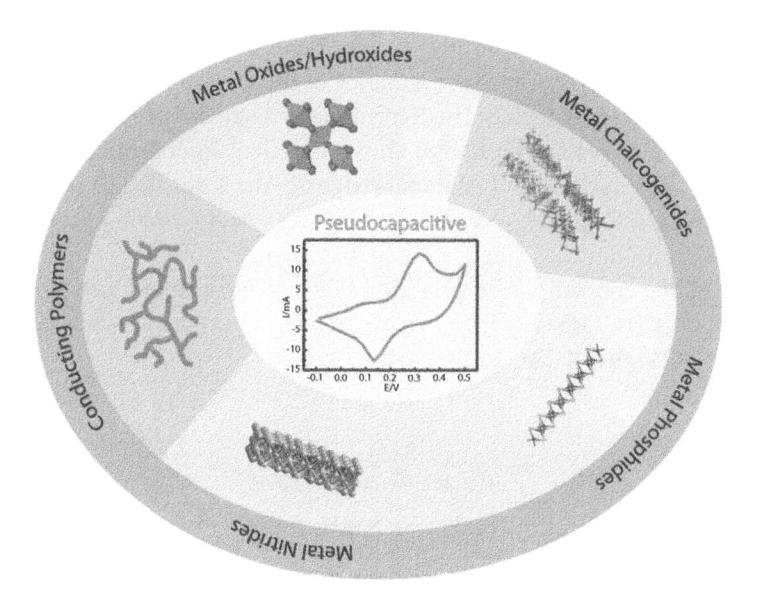

Figure 4.11 Schematic representation of different types of pseudocapacitive electrode materials.

capacitance. Generally, transition metal oxides/hydroxides/sulfides and conducting polymers are used as pseudocapacitive electrode materials. The different types of pseudocapacitive electrode materials are presented in Figure 4.11.

4.6.2.2.1   Metal Oxides/Hydroxides

Metal oxides and hydroxides based materials give more energy density to a supercapacitor as compared to EDLC materials and they have good electrochemical stability than the polymer-based materials. For energy storage performance of metal oxide based materials, these following points have been taken into consideration: (1) conducting/semiconducting nature of the material, (2) variable oxidation number of the metal atom, (3) electrochemical stability of the material during the ion intercalation and de-intercalation process. RuO_2 is the most extensively studied pseudocapacitive electrode material having three oxidation states accessible within wider potential range up to 1.2V. Hu et al. reported the highest value of 1300 F/g for hydrous RuO_2.[45] Similarly MnO_2 has been studied as a good candidate pseudocapacitive material due to its low cost, abundance and good electrochemical activity. MnO_2 give one electron transfer process Mn^{3+}/Mn^{4+} for the charge storage. Wang et al. reported 3D-α-MnO_2 nanostructures with a surface area of 284 m^2/g and give specific capacitance of 200 F/g.[46] Sahu et al. developed Mn3O4 hollow nanospheres on graphene surface and explored the energy storage performance of the electrode material at symmetric two electrode device. The hybrid material gives specific capacitance of 58 F/g at 1 mV/s scan rate and shows excellent operational stability of 97% over 2000 cycles.[47] Other metal oxides/hydroxides like Co_3O_4, Fe_2O_3, FeOOH Ni_3O_4, $Co(OH)_2$, $Ni(OH)_2$, VO_2, V_2O_5, MoO_3, WO_3 etc. with different morphology has been studied for energy storage application.[48-52] Kamila et al. reported a rare phase of $VO_2(D)$ and the reduced graphene oxide hybrid (rGO/VO_2(D)) as an electrode material for supercapacitor application (Figure 4.12).[53]

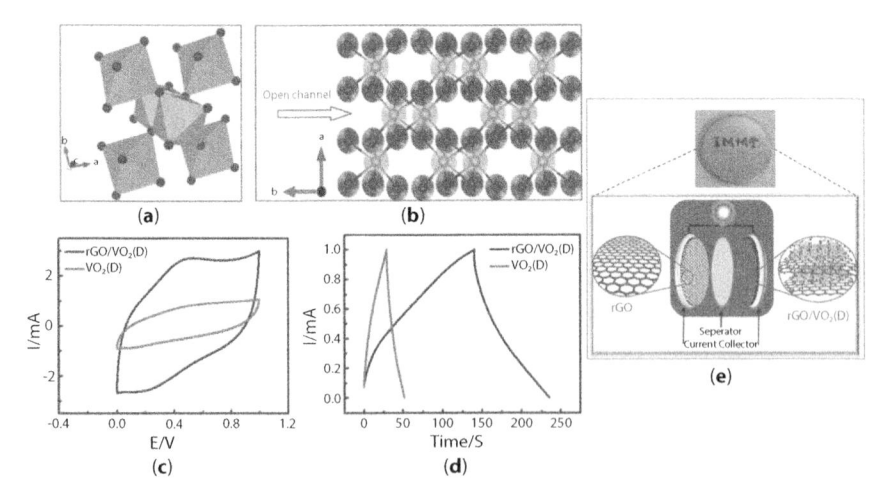

Figure 4.12 (a) DFT optimized structure of $VO_2(D)$, (c) overlap CV plot, (d) overlap CD plot of $VO_2(D)$ and rGO/$VO_2(D)$, (e) scheme for the fabrication of asymmetric coin cell device. [Reproduced with permission from ref 53].

In a symmetric supercapacitor device, the specific capacitance of rGO/VO_2(D) hybrid and bare VO_2(D) phase was found to be 737 F/g and 271 F/g at 1mV/s scan rate. Further, an asymmetric coin cell supercapacitor device has been fabricated and the electrochemical energy storage property tested at a wide working potential of 2V. The asymmetric device gives specific capacitance value of 147 F/g at 1mV/s scan rate. The energy density and power density performance were estimated and found to be 81 Wh kg^{-1}, 147 W kg^{-1} at 1 mV/s scan rate.[53] From the literature, it reveals that Co(OH)$_2$ materials give more capacitance value than that of Co_3O_4 nanomaterial. Zhou et al. established a mesoporous thin film of Co(OH)$_2$ on Ni foam having the specific capacitance of 2646 F/g.[54] Among the metal oxides, Nickel oxide materials give the high theoretical specific capacitance of 3750 F/g.[55] Layer double hydroxides are promising materials in the field of supercapacitor due to the facile tunability of its structure, composition and morphology. The interlayer spacing of LDH material helps for the increasing pseudocapacitance properties of the supercapacitor.[56]

4.6.2.2.2 Metal Chalcogenides

Recent research progress demonstrates that the transition metal chalcogenides are the group of a promising candidate for energy storage application due to their unique physical/chemical properties, intrinsic conductivity and electrochemical stability. The family of metal chalcogenides can be formed by disparate combination of metal (M) and chalcogenides (S/Se/Te) ratio. Those metal chalcogenides such as Nickel sulfides, cobalt sulfides, iron sulfides, copper sulfides and manganese sulfides with various morphology have been studied for supercapacitor application.[57] Beside these metal chalcogenides, some 2D layered transition metal dichalcogenides (TMDs) like MoS_2, $MoSe_2$, VS_2, VSe_2, WS_2, WSe_2, $NbSe_2$, $MoTe_2$ were investigated for supercapacitor performance. These TMDs have extraordinary properties like unique atomic and electronic arrangement, conducting and semiconducting in nature, large surface area, electrochemical stability over a wide potential window which makes it emerging in energy application. The graphene-like layered structure with interlayer spacing and the presence of variable oxidation state of metal atom allow these materials to store energy by both EDLC and pseudocapacitance mechanism.[58] Among them, MoS_2 is commonly used electrode material for supercapacitor application. Although VS_2, WS_2 has been used. MoS_2 gives theoretical specific capacitance of 1000 F/g. There are many research works reported for MoS_2 towards supercapacitor application. Typically, Ajayan and group developed MoS_2 based micro supercapacitor by spray painting

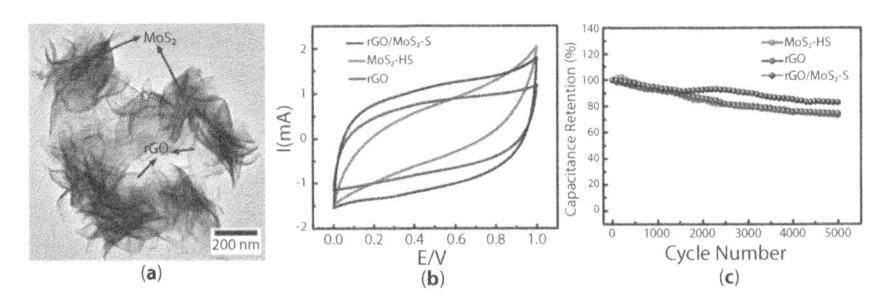

Figure 4.13 (a) TEM image of rGO/MoS$_2$-S, (b) overlap CV plot and (c) operational stability of MoS$_2$-HS, rGO and rGO/MoS$_2$-S [Reproduced with permission from ref 61].

process and subsequent laser patterning. This micro supercapacitor gives areal capacitance of 8 mF/cm^2 with excellent cycle life, that is better than the reported graphene-based microsupercapacitor.[59] Chhowalla et al. developed 1T phase of MoS$_2$ and studied the energy storage application.[60] Jena and groups synthesized MoS$_2$ hollow spheres and reduced graphene oxide hybrid (rGO/MoS$_2$-S) and reported the supercapacitor application.[61] The rGO/MoS$_2$-S shows gravimetric capacitance of 318 ± 14 Fg^{-1} at 1 mV/s scan rate and shows better cyclic performance of 82 ± 0.95% over 5000 cycles.[61]

4.6.2.2.3 Metal Nitrides/Phosphides

Like metal oxide/sulfides, transition metal nitrides and metal phosphides are eligible candidates for supercapacitor application. The good conductivity of the transition metal nitrides (4000 to 55500 S cm^{-1}) gives high power density performance compared to transition metal oxides. Also, the metal nitrides give good capacitance properties higher than the carbon-based materials and most of the metal oxides. There are so many works that have been done on a variety of transition metal nitrides and supercapacitor application.[62-65] Bouhitiyya et al. summarized various metal nitrides towards supercapacitor application in KOH electrolyte. It was found that VN gives highest specific capacitance as compared to other metal nitrides.[66] Choi et al. reported nanocrystalline high purity VN for supercapacitor performance. The energy storage mechanism of VN involves both EDLC and faradic reaction (II/IV) which gives a high specific capacitance value of 850 F/g at a scan rate of 50mV/s.[64] Apart from this chart, some other metal nitrides like molybdenum nitride, tungsten nitride, tantalum nitrides are reported for energy storage application.[52] Conway and co-workers established the energy storage behavior of the molybdenum nitride film are quite similar to RuO_2, which can be replaced due to high cost.[67] Kherani et al. reported than Mo_2N gives specific capacitance of 16 mF cm^{-2} which is 200 times more than its corresponding molybdenum oxide (0.67 mF cm^{-2}) in an acidic electrolyte.[68]

Metal phosphides are known to be n-type semiconductors and have both semi-metallic property and excellent conductivity. Compared to metal oxides/hydroxides, metal phosphides show fast reversible redox reaction, which gives high power density.[69] Nickel and Cobalt phosphides show rapid redox behavior with highly conducting and have high theoretical specific capacitance. Nickel phosphides contain different phases like $Ni_{12}P_{15}$, amorphous Ni-P, Ni_2P, Ni_4P_5 etc.[70-72] An et al. developed $Ni_{12}P_{15}$ hollow nano-capsules that exhibited better capacitance of 949 F/g at 1 A/g than those of NiO nanoflower (326 F/g) and $Ni(OH)_2$ of 787 F/g.[73] Zhou at al. synthesized Ni_2P on Ni foam that gives better capacitance of (3275 F/g at 2.5 A/g) i.e., three times higher than that of $Ni(OH)_2$/Ni foam electrode.[74] In the case of cobalt phosphides, the charge discharge process involves Co^{2+}/Co^{3+} state. There are few works that have been reported on cobalt Phosphides. Chen et al. have reported different morphology of cobalt phosphides like nano rod and nano flowers, and studied their energy storage performance.[75] Cobalt Phosphide nanoflower has high surface area, and gives better capacitance than nanorods.

4.6.2.2.4 Conducting Polymers

Conducting polymers are the potential pseudocapacitive materials of interest for energy storage application due to its low cost, good conductivity, high potential stability and good storage capacity. The commonly used polymers in supercapacitor are Polyaniline (PANI), Polypyrrole (Ppy), polythiophene and poly (3,4-ethylenedioxythiophene)-polystyrene sulphonate (PEDOT-PSS). There are so many works that have been established on conducting polymer-based materials for supercapacitor application. Among them, PANI is the most widely used electrode material in supercapacitor due to its high capacitance behaviour, good electronic conductivity and low cost. Pure PANI-based electrode has multiple redox reaction which gives specific capacitance of 600 F/g in aqueous electrolyte.[76] Wang and co-workers developed PANI nanowire by electrochemical polymerization process having high capacitance value of 950 F/g.[77] In the case of conducting polymers, the main disadvantage is the electrochemical stability during the ion intercalation and de-intercalation process. During the charge-discharge process, the swelling and shrinkage of the polymer backbone takes place, that causes the mechanical degradation of polymer chain. So, to avoid such limitation, composite materials have been developed by coupling polymers with other electrochemical stable materials like metal oxides/hydroxides and graphene.

4.6.3 Electrolytes

The electrolyte is the major component of a supercapacitor device. Three types of electrolytes are used in supercapacitor devices. i.e., aqueous, organic and ionic liquid. Acid- and alkali-based electrolytes have high ionic conductivity as compared to the organic electrolyte and give high power density. However, these devices have a narrow operating potential window of 1V due to the water decomposition which takes place at 1.23V.[78] It was reported that the operating potential can be extended up to 2V in an aqueous electrolyte-based asymmetric device by fabricating different nanostructured material at both cathode and anode with mass balancing.[79,80] It is important that the over potential of the electrolyte degradation can be avoided by using different carbon-based electrode material. It is reported that in neutral electrolytes like Na_2SO_4 and Li_2SO_4, the potential window is extended up to 2V by using activated carbon material in the symmetric device. Electrolytes composed of salt dissolved in the organic solvent has a large potential window of about 2.7-2.8V; however, the main limitation is the lower conductivity and low capacitance value.[81] But these high-voltage devices can give high energy density behavior, which makes it commercially viable. Propylene carbonate and acetonitrile are the most commonly used solvent in supercapacitor. By applying quaternary ammonium salts and tetraethyleneammonium tetrafluroborate (TEA-BF_4), the conductivity of electrolyte can be increased up to a saturation level. The ionic liquids are the molten salts at room temperature which are considered as a promising electrolyte for supercapacitor application.[82] Another advantage of this electrolyte is that it can be operated at an increased temperature range beyond 80 °C, which is a limitation for organic-based electrolyte. The ionic liquids are resistant to oxidation and reduction over a wide range of potential, providing cell voltage up to 4.5V - 6V. The main drawback of ionic liquid is lower conductivity less than 10 $mScm^{-1}$ which is significantly lower than aqueous electrolytes.[83] The high cost and the poor compatibility with the microporous carbon electrode material limits its commercial value.

4.6.4 Binders

Binders are important parts of a supercapacitor. The binder material is mixed with the active materials which provide good adhesion ability for the electrode material to the current collector as well as form a good electric network between the active material and conducting carbon for easy electron transportation and ion diffusion process. Generally, most commonly used fluorinated polymer-based binder materials are PVDF (polyvinylidene fluoride) and PTFE (polytetrafluoroethylene).[84] These polymers are chemical resistant to acid/alkaline-based electrolytes, having high electrochemical stability and good strength. PVDF is non-aqueous-based binder and it has good dispersibility in organic electrolyte, e.g., N-methyl-2-pyrrolidone. However, PTFE is compatible with both aqueous and organic mediums. Currently, carboxyl methyl cellulose (CMC), styrene butadiene rubber (SBR) are adopted as water-based binders in an electrochemical supercapacitor. To reduce the environmental concern, some other eco-friendly fluorine-free binders are developed such as polyacrylic acid (PAA), polyvinylidene pyrrolidone (PVP) and natural cellulose.[85-87] However, some conducting polymers like polyaniline (PANI), polypyrrole (PPY) and poly(3,4-ethylenedioxythiophene) or PEDOT are explored as a binder whose presence contributes capacitance behavior.[88,89] The performance of the supercapacitor device is affected by the amount and types of the binder used.[90,91] Because, the presence of a higher quantity of hydrophobic binder mitigates the electrolyte penetration and causes a decrease of capacitance. To avoid some negative impacts of the binder such as reduction of the active surface area, surface wettability, some works have been developed towards the binder-free electrode material towards supercapacitor application.

4.6.5 Separators

A separator is a part located between two electrodes that physically separates the electronic contact between cathode and anode but accesses the ions passing through it. There are so many parameters that should be kept in mind while choosing a separator for a supercapacitor device. The separator should be porous, chemically resistant towards the electrolyte, low resistance for the ion accessibility, high mechanical strength and flexibility, etc., which may help the device for better life span and good compatibility. Generally, porous thin films or membranes are used as a separator. The membranes are made up by cellulose-based materials, polymers and glass fibers. Glass fibers and some polymer membranes such as (polyvinyledene difloride (PVDF) and polytetrafluroethylene (PTFE) are used in alkaline electrolyte. Cellulose-based separators have good surface wettability in an aqueous medium and are compatible in neutral electrolyte. Although cellulose-based separator and polymer separators are used in an organic electrolyte, the cellulose-based separator suffers degradation in acidic medium.[92] Nafion@115, cellulose TF-40-30, celgard™, celgard polypropylene separators are used in acidic medium. The performance of the device is also dependent upon a different kind of separator taken.[93] Recently, some new separators have been explored like GO film and the egg shell membrane for supercapacitor device.[94]

4.7 Conclusion

Supercapacitors are considered to be a promising energy storage device and treated as ultra-native to batteries in the application where there are high power rates and fast energy storage demands. The nanostructure electrode materials play a major role in the device's better capacitance property due to their controllable size and morphology, shortening the ion diffusion path length and exposure of more outer surface area. Porous carbon nanomaterials, graphene and hetero atom doped graphene were proven to be good EDLC materials with high specific capacitance and useful in flexible and printable supercapacitor application. Generally, EDLC-based supercapacitors dominate the market over the other materials. Some pseudocapacitive electrode materials such as transition metal oxides/hydroxides/sulfides/nitrides/phosphides and conducting polymers are showing better energy storage property due to the ion intercalation process and surface redox reaction. However, the energy density performance of supercapacitor close to the rechargeable batteries is still challenging. So much research has been devoted to increasing the energy density performance of supercapacitors by using stable electrolytes with high working potential, developing of new nanostructure electrode materials having both capacitive type and battery type property, design of asymmetric supercapacitor, and fabrication of the battery-supercapacitor hybrid device. In this chapter, the basic concept of supercapacitors, covering charge storage mechanisms, electrode materials, and components of supercapacitors are summarized.

References

1. Hadjipaschalis, I., Poullikkas, A. & Efthimiou, V., Overview of current and future energy storage technologies for electric power applications. *Renewable and Sustainable Energy Reviews,* 13, 1513, 2009.
2. Lund, H., Renewable energy strategies for sustainable development. *Energy,* 32, 912, 2007.

3. Ibrahim, H., Ilinca, A. & Perron, J., Energy storage systems Characteristics and comparisons. *Renewable and Sustainable Energy Reviews,* 12, 1221, 2008.
4. Kittner, N., Lill, F. & Kammen, D., Energy storage deployment and innovation for the clean energy transition. *Nature Energy,* 2, 2017125, 2017.
5. Whittingham, M. S., Materials Challenges Facing Electrical Energy Storage. *MRS Bulletin,* 33, 411, 2008.
6. Nadeem, F., Hussain, S., Tiwari, P., Goswami, A. & Ustun, T. S., Comparative Review of Energy Storage Systems, Their Roles and Impacts on Future Power Systems. *IEEE Access,* 7, 4555, 2019.
7. Mandapati, J. & Balasubramanian, K., Simple Capacitors to Supercapacitors-An Overview. *Int. J. Electrochem. Sci* , 3, 1196, 2008.
8. Conway, B. E., Birss, V. & Wojtowicz, J., The role and utilization of pseudocapacitance for energy storage by supercapacitors. *Journal of Power Sources,* 66, 1, 1997.
9. Liang, Y., Zhao, C. Z., Yuan, H., Chen, Y., Zhang, W., Huang, J. Q., Yu, D., Liu, W., Titirici, M. M., Chueh, Y. l., Yu, H., Zhang, Q., A review of rechargeable batteries for portable electronic devices. *InfoMat,* 1, 6, 2019.
10. Libich, J., Máca, J., Vondrák, J., Čech, O. & Sedlaříková, M. Supercapacitors: Properties and applications. *Journal of Energy Storage,* 17, 224, 2018.
11. Wang, G., Zhang, L. & Zhang, J., A review of electrode materials for electrochemical supercapacitors. *Chem. Soc. Rev.,* 41, 797, 2012.
12. https://www.global.tdk.com/techmag/electronics_primer/vol8.html
13. https://www.prnewswire.com/news-releases/the-global-supercapacitor-market-is-expected-to-reach-31-billion-by-2026-300471006.html
14. https://www.lucintel.com/supercapacitor-market-2017-2026.aspx
15. Shttps://www.techsciresearch.com/report/india-supercapacitors-market/1370.html
16. Salele Iro, Z., A Brief Review on Electrode Materials for Supercapacitor. *International Journal of Electrochemical Science,* 11, 10628, 2016.
17. Pilon, L., Wang, H. & D'Entremont, A., Recent Advances in Continuum Modeling of Interfacial and Transport Phenomena in Electric Double Layer Capacitors. *Journal of the Electrochemical Society,* 162, A5158, 2015.
18. Burt, R., Birkett, G. & Zhao, X. S., A review of molecular modelling of electric double layer capacitors. *Physical Chemistry Chemical Physics,* 16, 6519, 2014.
19. Balasubramaniam, S., Mohanty, A., Balasingam, S. K., Kim, S. J. & Ramadoss, A., Comprehensive Insight into the Mechanism, Material Selection and Performance Evaluation of Supercapatteries. *Nano-Micro Letters,* 12, 85, 2020.
20. Kate, R. S., Khalate, S. A. & Deokate, R. J., Overview of nanostructured metal oxides and pure nickel oxide (NiO) electrodes for supercapacitors: A review. *Journal of Alloys and Compounds* 734, 89, 2018.
21. Zhao, B., Chen, D., Xiong, X., Song, B., Hu, R., Zhang, Q., Rainwater, B. H., Waller, G. H., Zhen, D., Ding, Y., Chen, Y.,Qu, C., Dang, D., Wong, C. P.,

Liu, M., A high-energy, long cycle-life hybrid supercapacitor based on graphene composite electrodes. *Energy Storage Materials,* 7, 32, 2017.
22. Gao, Y., Graphene and Polymer Composites for Supercapacitor Applications: a Review. *Nanoscale Research Letters,* 12, 387, 2017.
23. Barzegar, F., Momodu, D., Zhang, L., Xia, X. & Manyala, N., Design and characterization of asymmetric supercapacitor useful in hybrid energy storage systems for electric vehicles. *IFAC-PapersOnLine,* 50, 83, 2017.
24. Choudhary, N., Li, C., Moore, J., Nagaiha, N., Zhai, L., Jung. Y., Thomas, J., Asymmetric Supercapacitor Electrodes and Devices. *Advanced Materials,* 29, 1605336, 2017.
25. Zuo, W., Li, R., Zhou, C., Li, Y., Xia, J., Liu, J., Battery-Supercapacitor Hybrid Devices: Recent Progress and Future Prospects. *Advanced Science,* 4, 1600539, 2017.
26. Liu, C.-C., Tsai, D. S., Susanti, D., Yeh, W. C., Huang, Y. S., Liu, F. J., Planar ultracapacitors of miniature interdigital electrode loaded with hydrous RuO2 and RuO2 nanorods. *Electrochimica Acta,* 55, 5768, 2010.
27. Sumboja, A., Wang, X., Yan, J., Lee, P. S., Nanoarchitectured current collector for high rate capability of polyaniline based supercapacitor electrode. *Electrochimica Acta,* 65, 190, 2012.
28. Ahn, H.-J., Kim, W. B. & Seong, T. Y., Co(OH)2-combined carbon-nanotube array electrodes for high-performance micro-electrochemical capacitors. *Electrochemistry Communications,* 10, 1284, 2008.
29. Zhong, C., Deng. Y., Hu, W., Qiao, J., Zhang, L., Zhang, J., A review of electrolyte materials and compositions for electrochemical supercapacitors. *Chemical Society Reviews,* 44, 7484, 2015.
30. Ratajczak, P., Jurewicz, K., Béguin, F., Factors contributing to ageing of high voltage carbon/carbon supercapacitors in salt aqueous electrolyte. *Journal of Applied Electrochemistry,* 44, 475, 2014.
31. Su, Z., Yang, C., Xie, B., Lin, Z., Zhang, Z., Liu, J., Li, B., Kang, F., Wong, C. P., Scalable fabrication of MnO2 nanostructure deposited on free-standing Ni nanocone arrays for ultrathin, flexible, high-performance micro-supercapacitor. *Energy & Environmental Science,* 7, 2652, 2014.
32. Xiao, J., Yang, S., Wan, L., Xiao, F., Wang, S., Corrigendum to "Electrodeposition of manganese oxide nanosheets on a continuous three-dimensional nickel porous scaffold for high performance electrochemical capacitors". *Journal of Power Sources,* 254, 360, 2014.
33. Bittner, A. M., Zhu, M., Yang, Y., Waibel, H. F., Konuma, M., Starke, U., Weber, C. J., Ageing of electrochemical double layer capacitors. *Journal of Power Sources,* 203, 262, 2012.
34. Kamila, S., Manikandan, K., Chakraborty, B., Jena, B. K., Role of Iodine on HI treated flexible reduced graphene oxide film for enhancement of supercapacitance properties: Experimental with insight from DFT Simulations. *New Journal of Chemistry,* 44, 1418–1425, 2020.

35. Candelaria, S. L., Shao, Y., Zhou, W., Li, X., Xiao, J., Zhang, J. G., Wang, Y., Liu, J., Li, J., Cao, G., Nanostructured carbon for energy storage and conversion. *Nano Energy,* 1, 195, 2012.
36. Frackowiak, E., Carbon materials for supercapacitor application. *Physical Chemistry Chemical Physics,* 9, 1774, 2007.
37. Picó, F., Rojo, J. M., Sanjuan, M. L., Anson, A., Benito, A. M., Callejas, A. M., Maser, W. K., Martinez, M.T., Single-Walled Carbon Nanotubes as Electrodes in Supercapacitors. *Journal of The Electrochemical Society,* 151, A831, 2004.
38. Frackowiak, E., Metenier, K., Bertagna, V., Beguin, F., Supercapacitor electrodes from multiwalled carbon nanotubes. *Applied Physics Letters,* 77, 2421, 2000.
39. Chen, J., Li, C., Shi, G., Graphene Materials for Electrochemical Capacitors. *The Journal of Physical Chemistry Letters,* 4, 1244, 2013.
40. Sun, Y., Wu, Q., Shi, G., Graphene based new energy materials. *Energy Environ. Sci.,* 4, 1113, 2011.
41. Li, Z., Bu, F., Wei, J., Yao, W., Wang, L., Chen, Z., Pan, D., Wu, M., Boosting the energy storage densities of supercapacitors by incorporating N-doped graphene quantum dots into cubic porous carbon. *Nanoscale,* 10, 22871, 2018.
42. Gong, X., Liu, G., Li, Y., Yu, D. Y. W. & Teoh, W. Y. Functionalized-Graphene Composites: Fabrication and Applications in Sustainable Energy and Environment. *Chemistry of Materials,* 28, 8082, 2016.
43. Zhang, Z., Lee, C. S., Zhang, W., Vertically Aligned Graphene Nanosheet Arrays: Synthesis, Properties and Applications in Electrochemical Energy Conversion and Storage. *Advanced Energy Materials,* 7, 1700678, 2017.
44. Samantara, A. K., Sahu, S. C., Ghosh, A. and Jena, B. K., Sandwiched graphene with nitrogen, sulphur co-doped CQDs: an efficient metal-free material for energy storage and conversion applications, *J. Mater. Chem. A,* 3, 16961–16970, 2015.
45. Howse, J. R., Topham, P., Crook, C. J., Gleeson, A. J., Bras, W., Jones, R. A. L., Ryan. A. J., Reciprocating Power Generation in a Chemically Driven Synthetic Muscle. *Nano Letters,* 6, 73, 2006.
46. Wang, Y.-T., Lu, A.-H., Zhang, H.-L. & Li, W.-C., Synthesis of Nanostructured Mesoporous Manganese Oxides with Three-Dimensional Frameworks and Their Application in Supercapacitors. *The Journal of Physical Chemistry C,* 115, 5413, 2011.
47. Das, S, K., Kamila, S., Satpati, B., Kandasamy, M., Chakraborty, B., Basu, S., and Jena, B. K., Hollow Mn_3O_4 Nanospheres on Graphene Matrix for Oxygen Reduction Reaction and Supercapacitance Applications: Experimental and Theoretical Insight. *Journal of Power Sources,* 471, 228465, 2020
48. Barik, R., Jena, B. K., Mohapatra, M., In-situ synthesis of flowery shape α-FeOOH/Fe2O3 nano particles and their phase dependant supercapacitive behaviour. *RSC Advances,* 2014, 4, 18827–18834.

49. Barik, R., Jena, B. K., Mohapatra M., Metal doped mesoporous FeOOH nanorods for high performance supercapacitor. *RSC Advances,* 2017, 7, 49083–49090.
50. Ratha, S., Samantara, A. K., Sinha, K. K., Gangan, A. S., Chakraborty, B., and Jena, B.K., Rout, C. S., Urea-assisted Room Temperature Stabilized Metastable β-NiMoO$_4$: Experimental and Theoretical Insights into its Unique Bi-functional Activity towards Oxygen Evolution and Supercapacitor. *ACS Appl. Mater. Interfaces,* 9, 9640–9653, 2017.
51. Samantara, A. K., Kamila, S., Ghosh, A., Jena, B. K., Highly Ordered 1D NiCo$_2$O$_4$ Nanorods on Graphene: An Efficient Dual-functional Hybrid Materials for Electrochemical Energy Conversion and Storage Applications. *Electrochimica Acta,* 264, 147–157, 2018
52. An, C., Zhang, Y., Guo, H. & Wang, Y. Metal oxide-based supercapacitors: progress and prospectives. *Nanoscale Adv.* 1, 4644–4658, 2019.
53. Kamila, S., Chakraborty, B., Basu, S., and Jena, B. K., Combined Experimental and Theoretical Insights into Energy Storage Applications of a VO$_2$(D)–Graphene Hybrid. *J. Phys. Chem. C,* 123, 24280–24288, 2019.
54. Zhou, W.-J., Xu, M.-W., Zhao, D.-D., Xu, C.-L. , Li, H.-L., Electrodeposition and characterization of ordered mesoporous cobalt hydroxide films on different substrates for supercapacitors. *Microporous and Mesoporous Materials,* 117, 55, 2009.
55. Castro, E. B., Real, S. G., Pinheiro Dick, L. F., Electrochemical characterization of porous nickel–cobalt oxide electrodes. *International Journal of Hydrogen Energy,* 29, 255, 2004.
56. Li, X., Du, D., Zhang, Y., Xing, W., Xue, Q., Yan, Z., Layered double hydroxides toward high-performance supercapacitors. *Journal of Materials Chemistry A,* 5, 15460, 2017.
57. Rui, X., Tan, H. & Yan, Q. Nanostructured metal sulfides for energy storage. *Nanoscale,* 6, 9889, 2014.
58. Chia, X., Eng, A. Y. S., Ambrosi, A., Tan, S. M., Pumera, M., Electrochemistry of Nanostructured Layered Transition-Metal Dichalcogenides. *Chemical Reviews,* 115, 11941, 2015.
59. Cao, L., Yang, S., Gao, W., Liu, Z., Gong, Y., Ma, L., Shi, G., Lei, S., Zhang, Y., Zhang, S., Vajtai, R., Ajayan, P. M.,Direct laser-patterned micro-supercapacitors from paintable MoS2 films. *Small,* 9, 2905, 2013.
60. Acerce, M., Voiry, D., Chhowalla, M. Metallic 1T phase MoS2 nanosheets as supercapacitor electrode materials. *Nature Nanotechnology,* 10, 313, 2015.
61. Kamila, S., Mohanty, B., Samantara, A. K., Guha, P., Ghosh, A., Jena, B. L., Satyam, P. V., Mishra, B. K., Jena, B. K., Highly Active 2D Layered MoS2-rGO Hybrids for Energy Conversion and Storage Applications. *Scientific Reports,* 7, 8378, 2017.
62. Balogun, M.-S. Qiu, W., Wang, W., Fang, P., Lu, X., Tong, Y., Recent advances in metal nitrides as high-performance electrode materials for energy storage devices. *Journal of Materials Chemistry A,* 3, 1364, 2015.

63. Jampani, P. H., Manivannan, A., Kumta, P. N., Advancing the Supercapacitor Materials and Technology Frontier for Improving Power Quality. *The Electrochemical Society Interface,* 19, 57, 2010.
64. Choi, D., Kumta, P. N., Chemically Synthesized Nanostructured VN for Pseudocapacitor Application. *Electrochemical and Solid-State Letters,* 8, A418, 2005.
65. Zhou, X., Chen, H., Shu, D., He, C. & Nan, J., Study on the electrochemical behavior of vanadium nitride as a promising supercapacitor material. *Journal of Physics and Chemistry of Solids,* 70, 495, 2009.
66. S. Bouhtiyya, R. Lucio-Porto, J.-B. Ducros, P. Boulet, F. Capon, T. Brousse, J. F. Pierson, Transition Metal Nitrides Thin Films for Supercapacitor Applications. *ECS Meeting Abstract,* MA2012-02, 494, 2012.
67. Liu, T. -C., Pell, W. G., Conway, B. E., Roberson, S. L., Behavior of Molybdenum Nitrides as Materials for Electrochemical Capacitors: Comparison with Ruthenium Oxide. *Journal of the Electrochemical Society,* 145, 1882, 1998.
68. Ting, Y.-J. (Bernie), Lian, K., Kherani, N., Fabrication of Titanium Nitride and Molybdenum Nitride for Supercapacitor Electrode Application. *ECS Transactions,* 35, 133, 2019.
69. Li, X., Elshahawy, A. M., Guan, C., Wang, J., Metal Phosphides and Phosphates-based Electrodes for Electrochemical Supercapacitors. *Small,* 13, 1701530, 2017.
70. Wang, X., Kim, H.-M., Xiao, Y., Sun, Y.-K., Nanostructured metal phosphide-based materials for electrochemical energy storage. *Journal of Materials Chemistry A,* 4, 14915, 2016.
71. Lu, Y., Liu, J. K., Huang, S., Wang, T. Q., Wang, X. L., Gu, C. D., Tu, J. P., Mao, S. X., Facile synthesis of Ni-coated Ni2P for supercapacitor applications. *CrystEngComm,* 15, 7071, 2013.
72. Wang, D., Kong, L. B., Liu, M. C., Zhang, W. B., Luo, Y. C., Kang, L., Amorphous Ni–P materials for high performance pseudocapacitors. *Journal of Power Sources,* 274, 1107, 2015.
73. Wan, H. et al. One pot synthesis of Ni12P5 hollow nanocapsules as efficient electrode materials for oxygen evolution reactions and supercapacitor applications. *Electrochimica Acta,* 229, 380, 2017.
74. Zhou, K., Zhou, W., Yang, L., Lu, J., Cheng, S., Mai, W., Tang, Z., Li, L., Chen, S., Ultrahigh-Performance Pseudocapacitor Electrodes Based on Transition Metal Phosphide Nanosheets Array via Phosphorization: A General and Effective Approach. *Advanced Functional Materials,* 25, 7530, 2015.
75. Chen, X., Cheng, M., Chen, D., Wang, R., Shape-Controlled Synthesis of Co2P Nanostructures and Their Application in Supercapacitors. *ACS Applied Materials & Interfaces,* 8, 3892, 2016.
76. Li, H., Wang J., Chu, Q., Wang, Z., Zhang, F., Wang, S., Theoretical and experimental specific capacitance of polyaniline in sulfuric acid. *Journal of Power Sources,* 190, 578, 2009.

77. Wang, K., Huang, J. & Wei, Z. Conducting Polyaniline Nanowire Arrays for High Performance Supercapacitors. *The Journal of Physical Chemistry C*, 114, 8062, 2010.
78. Ruiz, V., Blanco, C., Pinero, E.R., Khomenko, V., Beguin, F., Santamaria, R., Effects of thermal treatment of activated carbon on the electrochemical behaviour in supercapacitors. *Electrochimica Acta*, 52, 4969, 2007.
79. Singh, A., Chandra, A. Significant Performance Enhancement in Asymmetric Supercapacitors based on Metal Oxides, Carbon nanotubes and Neutral Aqueous Electrolyte. *Scientific Reports*, 5, 15551, 2015.
80. Chen, Y.-C., Lin, Y. G., Novel Iron Oxyhydroxide Lepidocrocite Nanosheet as Ultrahigh Power Density Anode Material for Asymmetric Supercapacitors. *Small*, 10, 3803, 2014.
81. Gamby, J., Taberna, P. L., Simon, P., Fauvarque, J. F. & Chesneau, M., Studies and characterisations of various activated carbons used for carbon/carbon supercapacitors. *Journal of Power Sources*, 101, 109, 2001.
82. Salanne, M. Ionic Liquids for Supercapacitor Applications. *Topics in Current Chemistry*, 375, 63, 2017.
83. Galiński, M., Lewandowski, A. & Stępniak, I. Ionic liquids as electrolytes. *Electrochimica Acta* 51, 5567, 2006.
84. González, A., Goikolea, E., Barrena, J. A. & Mysyk, R., Review on supercapacitors: Technologies and materials. *Renewable and Sustainable Energy Reviews*, 58, 1189, 2016.
85. Zhu, Z., Effects of Various Binders on Supercapacitor Performances. *International Journal of Electrochemical Science*, 11, 8270, 2016.
86. Böckenfeld, N., Jeong, S. S., Winter, M., Passerini, S., Balducci, A., Natural, cheap and environmentally friendly binder for supercapacitors. *Journal of Power Sources*, 221, 14, 2013.
87. Lee, K. T., Tsai, C. B., Ho, W. H. Wu, N.-L., Superabsorbent polymer binder for achieving MnO2 supercapacitors of greatly enhanced capacitance density. *Electrochemistry Communications*, 12, 886, 2010.
88. Aslan, M., Weingarth, D., Jackel, N., Atchison, J. S., Grobelesek, I., Presser, V., Polyvinylpyrrolidone as binder for castable supercapacitor electrodes with high electrochemical performance in organic electrolytes. *Journal of Power Sources*, 266, 374, 2014.
89. Lee, H., Kim, H., Cho, M. S., Choi, J., Lee, Y., Fabrication of polypyrrole (PPy)/carbon nanotube (CNT) composite electrode on ceramic fabric for supercapacitor applications. *Electrochimica Acta*, 56, 7460, 2011.
90. Lei, C., Wilson, P., Lekakou, C., Effect of poly(3,4-ethylenedioxythiophene) (PEDOT) in carbon-based composite electrodes for electrochemical supercapacitors. *Journal of Power Sources*, 196, 7823, 2011.
91. Abbas, Q., Pajak, D., Frąckowiak, E., Beguin, F., Effect of binder on the performance of carbon/carbon symmetric capacitors in salt aqueous electrolyte. *Electrochimica Acta, 140, 132*, 2014

92. Tsay, K.-C., Zhang, L., Zhang, J., Effects of electrode layer composition/thickness and electrolyte concentration on both specific capacitance and energy density of supercapacitor. *Electrochimica Acta*, 60, 428, 2012.
93. Liu, X., Pickup, P. G., Performance and low temperature behaviour of hydrous ruthenium oxide supercapacitors with improved power densities. *Energy & Environmental Science*, 1, 494, 2008.
94. Shulga, Y. M., Baskakov, S. A., Smirnov, V. A., Shulga, N. Y., Belay, K.G., Gutsev, G.L., Graphene oxide films as separators of polyaniline-based supercapacitors. *Journal of Power Sources*, 245, 33, 2014.

5

Thermal Energy Storage Systems for Cooling and Heating Applications

Pankaj Kalita[1*], Debangsu Kashyap[1] and Urbashi Bordoloi[2]

[1]*Centre for Energy, Indian Institute of Technology Guwahati, Guwahati, Assam*
[2]*Centre for Rural Technology, Indian Institute of Technology Guwahati, Guwahati, Assam*

Abstract

This chapter focuses on the importance of Thermal Energy Storage (TES) technology and provides a state-of-the-art review of its significance in the field of space heating and cooling applications. The chapter starts with a brief introduction followed by the classification of different commonly used TES technologies, viz. sensible heat storage (SHS), latent heat storage (LHS), and thermochemical heat storage (TCHS). A major portion of the chapter discusses the importance of these three systems and their real-time application and technological interventions. A systematic analysis has been made for both passive and active systems on space heating and cooling of both residential and commercial buildings. The active system of space heating includes solar FPS, solar air heater, solar pump, solar PV/T and solar pond. Different passive architecture systems and the use of phase change materials in passive buildings have also been explained. Space cooling achieved through solar refrigerating systems, viz. vapour adsorption/absorption system is analyzed, and detailed comparisons on the advantages and limitations of different TES systems are discussed. Further, sufficient numbers of numerical and design problems have been incorporated to provide a comprehensive understanding and design of the TES system for space heating/cooling applications.

Keywords: Energy storage, sensible heat, latent heat, thermochemical, heating, cooling

---

Corresponding author: pankajk@iitg.ac.in

Umakanta Sahoo (ed.) Energy Storage, (149–200) © 2021 Scrivener Publishing LLC

5.1 Introduction

In the present world, judicial energy utilization and its application play a crucial role in the technological advancement and economic growth of a country. The primary energy consumption is likely to rise by up to 48% by 2040, and hence renewable energy sources, waste energy and high technology systems will be the prominent future energy generation systems. These energy systems will not only provide sustainable energy for production and transportation but will also limit the overexploitation of fossil fuels. According to Faraj *et al.* building sectors are one of the largest energy-consuming sectors that account for more than 33.3% of total energy consumption globally and are also one of the major contributors to CO_2 emission. With more than half of the energy consumption of a building coming from ventilation, air conditioning, and heating systems, the conventional systems need to be replaced with non-conventional energy systems. Renewable energy systems such as solar, biomass, hydro and wind are playing a significant role as far as growing energy demand is concerned. Among these renewable systems, solar energy utilization and its applications stand out due to its enormous potential, cleanliness and natural availability. Solar energy finds its application in different fields, viz. electricity generation, refrigeration system, space cooling and heating systems. However, due to the random and intermittent supply of solar energy, the regular supply of energy is hampered. In particular, meeting the cooling and heating demand of buildings and providing continuous energy supply daily and seasonally has been a major issue among governments and societies. The aim is to provide reduced energy consumption without affecting the comfort conditions of the end user. Hence a novel technology needs to be developed for providing efficient energy and meeting the energy consumption in buildings. Integrated Thermal energy storage (TES) has been proven as a suitable solution, allowing the integration of renewable energy systems with TES for providing efficient energy consumption. The idea of the TES system was first mentioned and investigated in 1970 to address the energy shortage crisis. The primary function of a TES system is to stock up energy during heating and cooling of storage medium for utilizing the stored energy as required for heating and cooling applications in later stages. A TES system used in space cooling and heating systems aims at bridging the gap between demand and supply, enhancing both flexibility and performance of the renewable energy integrated system into thermal networks and also providing energy efficiency in buildings. One such TES system in Europe has been estimated to save around 1.4 million GWh/year and reduction in 400 million tons of CO_2 emission when utilized for heat and cold storages. Different energy storage technologies utilize a temperature range of -40 °C to more than 400°C as sensible heat, latent heat and thermochemical energy.

This chapter focuses on different TES technologies that utilize primarily solar energy for space heating/cooling applications and also meeting the energy demand in buildings. The working principles of different energy storage systems and their application in diverse fields are discussed in this chapter. Sensible heat storage systems (SHS), which include water, packed bed, aquifers, latent heat storage system (LHS), which mostly uses PCM materials and thermochemical heat storage system (TCHS), which includes absorption and adsorption systems along with their comparative analysis, are critically discussed in this chapter.

5.2 Classification of Storage Systems

The intermittency in a constant and continuous supply of solar radiations has encouraged many researchers towards the direction of energy storage systems. Different types of TES systems are developed based on the requirement and application for improving the performance and thermal reliability of any thermal system. Designing of a suitable TES system involves the characterization of several parameters, viz. capacity of the system, power requirement, efficiency, storage period, charging and discharging time and the cost of the system. Based on these characteristics, the TES system is mainly classified into three systems, as shown in Figure 5.1, viz. sensible storage system, latent storage system and chemical storage system. As far as water heating is considered, sensible heat storage is logical; for air heating collectors, sensible and latent heat storage systems are suitable. For systems involving photovoltaic and photochemical processes, chemical storage is sensible.

5.3 Sensible Heat Storage

Sensible heat storage uses the technique of storing heat (cold) with an increase or decrease in temperature of the storage medium. In other words, the heat capacity and the temperature difference of the storage medium is utilized during the charging and discharging processes. It is one of the simplest and currently the most developed TES systems which is commercially available compared to other TES systems. SHS systems mainly consist of

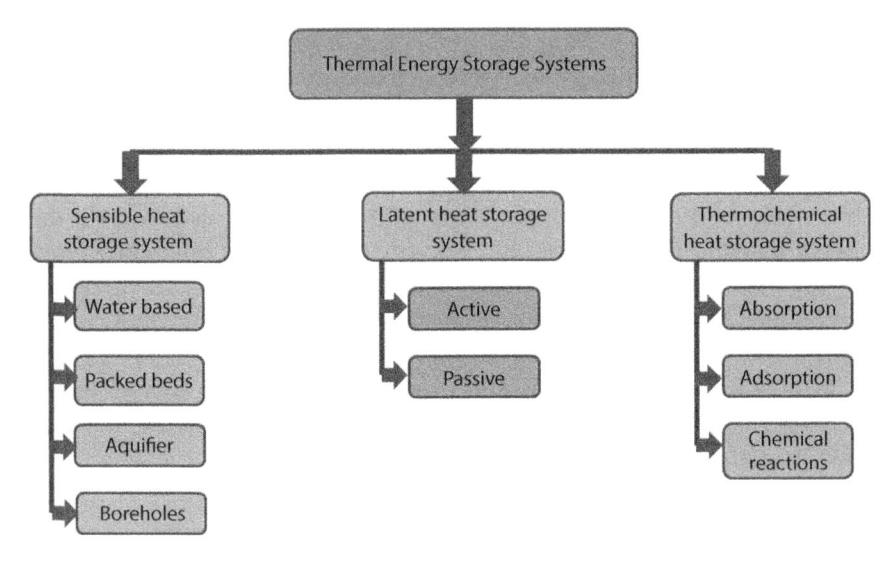

Figure 5.1 Classification of different TES systems.

a container, a storage medium and an input/output device. The amount of heat energy (Q_s) stored or released in SHS is expressed as

$$Q_S = \int_{T_i}^{T_f} mc_p \Delta T = mc_p(T_f - T_i) \tag{5.1}$$

Where Q_s is the amount of heat stored in Joules (J), m is the mass of heat storage medium in kg, c_p is the specific heat (J/kg/k), T_f and T_i are the initial and final temperatures of the storage material (K).

Here specific heat (c_p) is a temperature-dependent entity. However, for small variations in temperature range, approximate results can be obtained at constant c_p values for which Eq. (5.1) can be rewritten as

$$Q_S = m\bar{c}_p(T_f - T_i) = \rho V \bar{c}_p(T_f - T_i) \tag{5.2}$$

where $\bar{c}_p$ is the average specific heat of the material for a particular range of temperature. ρ is the density (kg/m3) of storage material which is considered almost constant for most of the storage materials foe building applications. However, for packed bed or porous medium storage material, the porosity of the medium has to be taken into account for which energy density per unit mass or volume for certain materials can be calculated respectively as Eq. (5.3) and Eq. (5.4)

$$\frac{Q_s}{m} = \bar{c}_p(T_f - T_i) \quad (5.3)$$

$$\frac{Q_s}{V} = \rho \bar{c}_p(T_f - T_i) \quad (5.4)$$

From Eq. (5.3) and Eq. (5.4) it can be concluded that higher values of c_p depict a correspondingly high energy storage density and vice-versa. Hence c_p can be considered to be a key parameter in the selection of suitable SHS material.

Some of the other parameters required to be considered for the selection of sensible heat storage systems are discussed in Table 5.1.

It should be noted that all the above-described properties may not be necessarily be fulfilled by a sensible heat storage system, and mostly the mentioned properties are interdependent. For example, for increasing the storage capacity of the system, thicker insulating layers can be used, but this will impact the size and capital cost of the system.

5.3.1 Water-Based Storage

Water-based storages which utilize water as the storage medium or as a heat carrier fluid is considered to be a favourable material for energy storage due to its higher specific heat in comparison with other sensible heat storage material. In addition, a higher capacity rate of water is also favourable while being charged or discharged. The sources of power for such water tanks could be variable; conventional gas/electric boilers, combined heat and power (CHP), air-water heat pumps integrated with water storage systems are the commonly used systems in buildings across the globe. Moreover, renewable energy systems, viz. solar flat plate collectors (FPC), solar PV panels, solar vacuum tube collectors, fuel cells or geothermal energy can also be used as a power source for water-based systems. A typical water storage system is shown in Figure 5.2.

Water tanks can be used for the storage of both hot and cold water. Depending on the heat source of the power supply, usually hot water tank stores water from about 40 °C to 80 °C, which is utilized for space heating and domestic hot water production. The stored hot water with the help of absorption refrigeration systems or thermoelectric elements can be used

Table 5.1 Properties of SHS materials.

Properties	Desirable characteristics
High Energy Density	Higher storage density requires a lower amount of storage material for certain materials resulting in a lower capacity cost of the storage system.
Higher energy efficiency	Higher energy efficiency requires a high proportion of energy storage in the system. This can be achieved primarily by reducing energy loss during charging and discharging. However, factors such as temperature difference between stored medium and environment, storage duration, insulation, etc., also contribute to the amount of energy loss.
Wide operation range of temperature	For a particular application-wide temperature range is favourable without phase change or decomposition of storage materials. Especially for building applications, working temperature ranges from 0-120°C, except for specific purposes.
Fast charging and discharging	High thermal conductivity, high heat capacity, Long-term thermal cycling stability properties are required for reducing the time to reach the storage capacity, also, for efficient heat transfers inside the storage medium as well as between the storage medium and heat transfer fluids.
Mechanical properties	Good mechanical stability for short-term and long-term storage, which means low degradation and stable thermo-physical properties of storage material.
Environmental	Storage material should be low or non-corrosive, environment-friendly, the lower energy requirement for manufacturing and CO_2 footprint.
Economic properties	Lower cost of manufacturing which refers to either low capacity cost or low power cost of storage material.

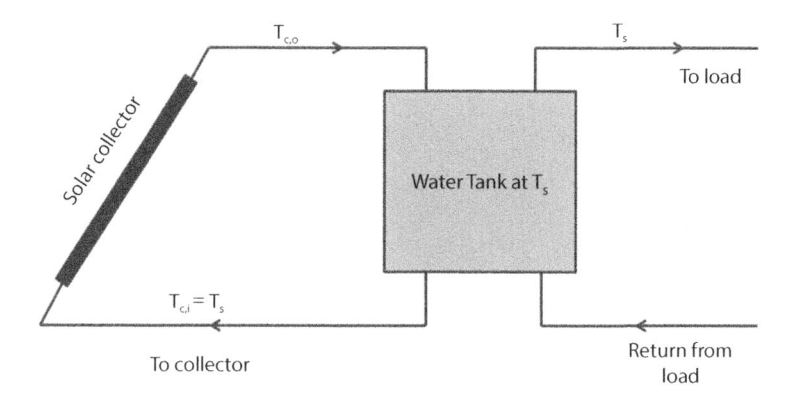

Figure 5.2 Schematic of a typical water tank storage system.

for space cooling downstream of the water tank. On the other hand, in cold water tanks, chilled water can be stored usually at a temperature range of 3 °C to 15 °C for the purpose of space cooling.

Water tank location and material play a key role in storage efficiency. Such tanks can be located either inside or outside or even underground of buildings with different sizes (as per requirement for a single room or district heating and cooling plants) and geometries, viz. vertical/horizontal either rectangular or cylindrical. Water tanks can be made of a wide variety of materials, which include aluminium, steel or reinforced concrete. For making the water tank lighter in weight, advanced materials such as fibre/plastic composites, encasement material and expanded polystyrene could be used as storage material, which is corrosion free and reduces the weight of the tank to almost one-third as compared to a tank made of steel. Another factor that contributes to an increase in efficiency is proper insulation of the tank to avoid thermal losses to the ambient. Conventional materials such as mineral wool, glass wool, polyethylene terephthalate-fibre fleeces, or eco-cotton wool greatly help reducing energy losses. Lottner *et al.* employed textile bags filled with granulated foam glass between soil and high-density concrete wall of artificial tanks. Researchers of Technique University of Ilmenau used glass fibre reinforced plastics as wall material along with an integrated heat insulation layer. In addition, advanced insulating materials such as silica aero gel-based or vacuum insulation panels having very lower thermal loss rate is one of today's prominent research activities.

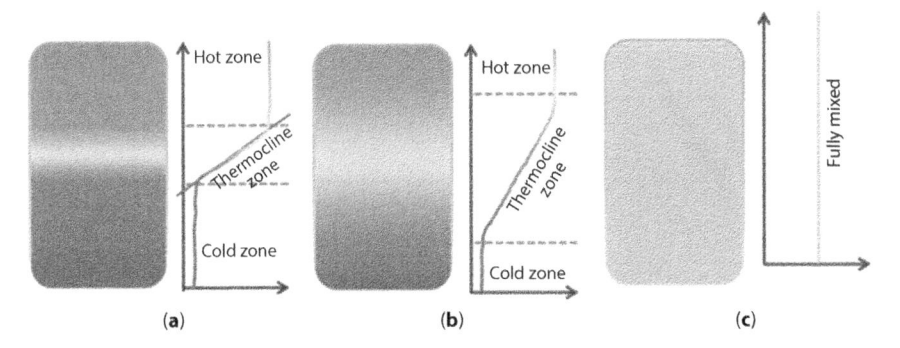

Figure 5.3 Different degree of stratification with same amount of stored heat within a storage tank; (a) highly stratified, (b) moderately stratified and (c) fully mixed, unstratified storage.

The water storage tank operates through the process of stratification. In this process, due to thermal buoyancy, the water at the top of the tank is hotter than at the bottom of the tank, and the consequent mixing effect due to temperature difference degrades the heat source level and negatively impacts the efficiency of the system. However, Ghaddar found that using a fully stratified tank instead of a fully mixed one can increase the energy storage efficiency up to 20% in solar water heating applications. Han *et al.* developed a horizontally portioned water tank with thermal stratification at SJTU. The results indicated a temperature difference of 70 K between the inlet and outlet of the tank and a temperature gap of 15-20 K between each chamber. An improved and better thermal stratification could be achieved with the horizontal partitioned tank resulting in efficient heat storage application. The effects of different degrees of stratification of a water storage tank of the constant volume are shown in Figure 5.3. Some of the other methods include optimization of geometrical parameters, viz. tank size, height to diameter ratio, selection of appropriate operating conditions, viz. temperature, inlet flow velocity, cycle duration, etc. In addition, researchers have also proposed multiple charging/discharging instead of conventional two levels of charging/discharging where the intermediate level is introduced at variable height in the middle of the tank resulting in simultaneous charging/discharging at different temperature levels without much disturbance of temperature stratification.

5.3.2 Packed Beds

Packed beds or simply rock bed heat storage system utilizes pebble, gravel or bricks beds circulated with heat transfer fluid for exchanging heat.

Packed bed systems can endure much higher temperatures compared to water-based systems. However, packed-bed systems require much larger volumes due to its low energy density almost three times as compared to water-based systems. In packed-bed systems, the tank is usually buried underground, insulated on sides and top to prevent thermal losses. Heat transfer fluid is circulated through pipes installed at different layers to release or absorb heat. Packed-bed systems find their place in large-scale application in district cooling/heating plants because of the cheaper costs of the storage medium and envelope structure. In building application, packed beds can be collectively used with solar collectors either with direct integrations or indirectly through pipes and ducts connection of the tank with solar collectors. The heat from the solar collector passes through the packed bed from top to bottom during the charging phase to release the heat. On the other hand, during the discharging phase, air from the building passes from the bottom to top of the bed, absorbing the stored heat and finally delivering the heated air into the building. Such type of space heating was performed by Zhao *et al.* on Qinhuang Island, China, as shown in Figure 5.4 where a 717 m^2 dormitories and a 2602 m^2 cafeteria were used for solar air heating with durations of 24h and 5 h, respectively. The solar collector of 473.2 m^2, which was used as a heat source, was used, and the surplus heat collected during daytime was stored in a 300m^2 pebble beds, and stored heat was utilized during night. Air was used as the heat transfer fluid as it could easily use the stored heat for indoor heat and also cost-effective. The experimental results revealed that the heating system was able to provide heat to the indoor areas and also store heat at the same time with a mean solar fraction of 19.1% and the highest value of 33.6%.

Thermal performance and pressure drop of the packed bed storage unit are greatly influenced by the size of packing materials and void fraction. As reported by SORUR, intermediate particle size and lower flow rates of HTF are advantageous in improving the thermal efficiencies of the storage unit. A trade-off between thermal performance and pressure drop is very much essential in designing of packed-bed storage units. In addition to these conventional concepts, some new technology interventions have been proposed by different researchers to intensity the heat transfer. These include the use of storage bins with trays instead of randomly packed rocks to carry rocks as proposed by AUDI. Another researcher proposed the use of cascade storage tanks instead of a single tank and more of such coupled or combined rock bed/water-based and rock bed/solar pond systems.

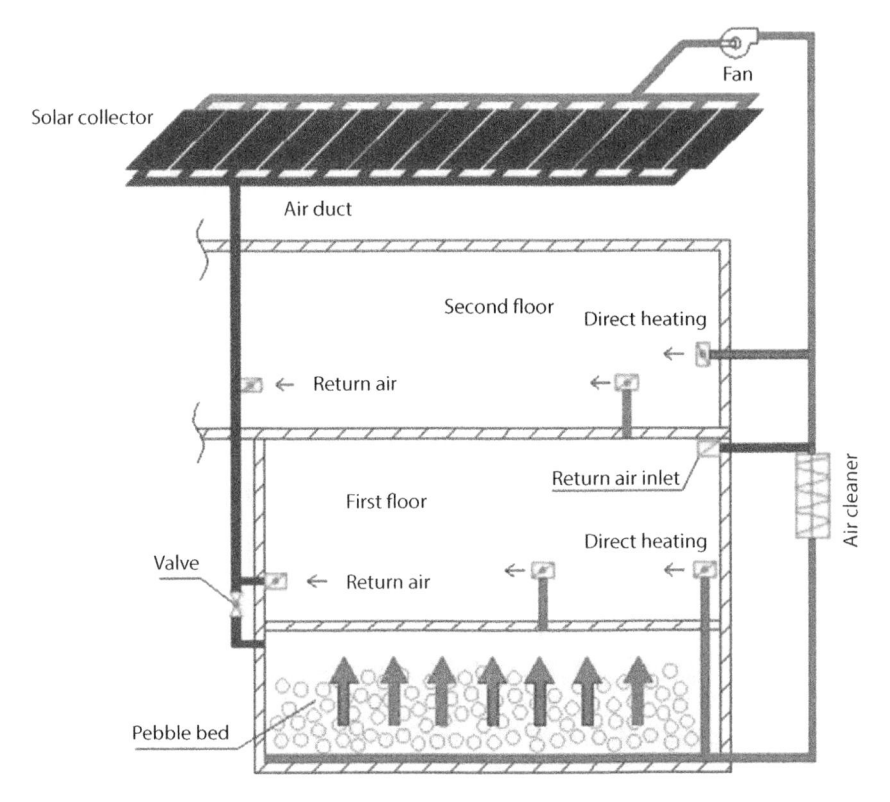

Figure 5.4 Schematic of a solar air heating system integrated with packed bed thermal storage system.

5.3.3 Aquifers

Aquifers are similar to packed-bed storage systems except for the fact that water is the primary storage medium, which flows at low rates through the ground. Water flow acts as the heat exchanging medium with the ground as it is pumped in and out of the ground to heat it and extract heat from it. Aquifers are natural water tanks and are considered as an effective cost-effective option as it limits the investments on underground construction and evacuation. Due to higher specific heat capacity and porous media of water, aquifers serve a proficient medium where heat can be stored and retrieved. In such systems saturation zone of groundwater is used for storage purposes. The aquifer consists of two thermal wells, viz. hot well and cold well drilled into it. A schematic representation of an aquifer system in shown in Figure 5.5. During the summer season, the heat source heats up the cold water extracted from the cold tank and rejects into the warm

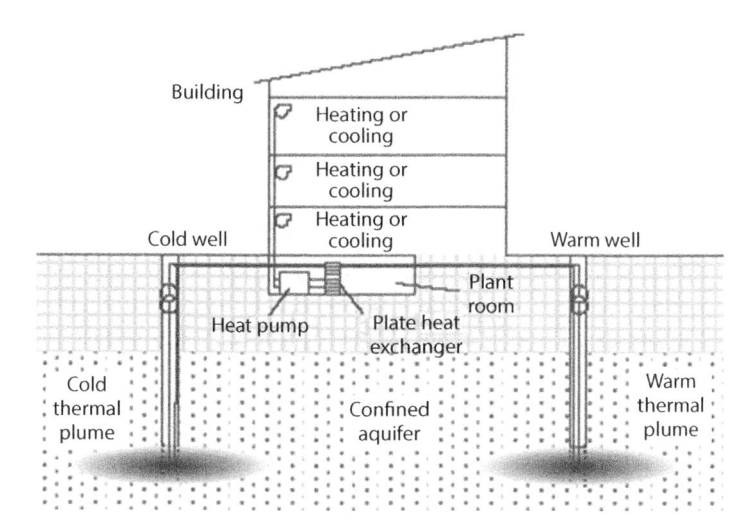

Figure 5.5 Schematic of a typical ATES system.

well; and during the winter season, the process gets reversed, i.e., hot water extracted from the warm well is cooled and rejected to the cold well. Such type of systems is usually preferred for large-scale district heating and cooling.

The first of its kind aquifer TES system was launched in 1976 and has been in continuous use thereafter. The aquifer TES system developed by Schmidt *et al.* in Germany showed promising working efficiency. The system was 30 m deep with an operating temperature ranging between 10-50 °C and was coupled with a solar collector of size 1000 m^2. ATES system was used space heating and supplying hot water for 108 apartments with a heated area of 7000 m^2. The long-term operation of the system resulted in a solar fraction of 62%, and also the solar energy captured during summertime was stored for compensating the energy storage during the winter season. Another ATES system in a hospital near Antwerp, as reported by Vanhoudt *et al.* consisting of two wells, was utilized for heating, cooling and also for regeneration mode. The system provided about 81% of total cooling energy, 22% of heating from the groundwater, saving a total of 1280 tons of CO_2.

For the ATES system to be feasible, some of the geological conditions play a decisive role. These factors include higher ground porosity, suitable chemical stability between groundwater and the matrix, limited groundwater flow through reservoir, etc. In addition, for the prevention of scaling and well-clogging, good knowledge of minerals and ground chemistry is essential.

5.3.4 Borehole

Boreholes are a type of TES system where the ground itself is directly used as the storage material. In this kind of storage approach, vertical or horizontal tubes are erected on the excavated and drilled ground where the underground structure is used for storing large amounts of solar heat collected during the summer for use in winters. The heat from the ground is absorbed through the HTF, usually glycol or water circulated through the tubes. Generally, the temperature difference between HTF and ground is small, and hence for improving the efficiency, it is recommended to use borehole TES in combination with heat pumps. The borehole heat exchangers are of different types, which include concentric-pipe, U-pipe or double U-pipe.

The BTES system has received considerable attention of large-scale seasonal plants due to its adaptive feature. Water clay or clay stones are one of the suitable BTES systems having higher heat capacities and preventing groundwater flow. However, due to lower energy storage density the system requires 3-5 times more volume compared to a water-based system for the same amount of stored energy. Again in comparison with Aquifer TES, the BTES system has a higher initial cost and takes much longer time to reach suitable performance increases the estimated payback time for the system. Another important issue of the BTES system is the controlled underground transfer of heat. The thermal losses should be reduced, and the heat transfer between the HTF inside the tube and surrounding ground should be enhanced. This includes the improvement in thermal properties of the material, suitable contact between the ground and the tubes, geochemical condition of the location, arrangement of tubes underground, favourable groundwater, thermal conductivity, etc.

A BTES system with a crystalline rock in Stockholm was the first of its kind borehole storage system (Lundh and Dalenback). A hundred boreholes of 65m depth with double U-pipes configuration coupled with 2400 m^2 solar collectors. An average of 70% solar fraction was obtained after 3-5 years of operation. An integrated large seasonal storage BTES system was developed by Drake landing Solar Community in Okotoks, Canada. The system consists of a BTES system for seasonal storage, solar collection, an Energy centre for short-term energy storage, a district heating system and energy-efficient houses. Boreholes of 35 m depth as many as 144 were drilled, covering an area of 35m diameter underground. The integrated system supplied 80% of the total energy demand of the entire community in winter and proved the feasibility of a BTES storage system at a higher latitude district.

Comparison of different Sensible Thermal Energy storage systems is figured out in Table 5.2.

TES SYSTEMS FOR COOLING AND HEATING APPLICATIONS 161

Table 5.2 Comparison of different Sensible Thermal Energy storage systems.

STES system	Storage medium	Geological requirements	Storage capacity (kWh/m³)	Advantages	Disadvantages
Water based	Water	▲ Depth 5-15m ▲ Stable ground conditions. ▲ Preferably no water	60-80	▲ Most common system ▲ Suitable for almost any location ▲ High stratification ▲ Can be used in any geological conditions.	▲ Higher thermal loss ▲ Higher cost ▲ Leakage and corrosion
Packed beds	Water and Gravel	▲ Heat transfer fluid. ▲ Insulated on top and sides	30-50	▲ Can be built almost anywhere ▲ More cost effective than water tank ▲ Any geological condition is satisfied	▲ Lower energy density so larger storage volume requirement. ▲ Higher cost ▲ Lower thermal conductivity results lower stratification.
Aquifer	Water-sand/gravel	▲ Natural water tank with high thermal conductivity ▲ Minimum or zero ground water flow ▲ 20m to 50m depth aquifer thickness	30-40	▲ Suitable for both heating/cooling ▲ Low maintenance cost ▲ Compared to borehole it is much more efficient ▲ Cost effective	▲ Requirement of special geological conditions ▲ Larger volume of tank (2-3 times) compared to water storage system ▲ Higher thermal loss

(*Continued*)

Table 5.2 Comparison of different Sensible Thermal Energy storage systems. (*Continued*)

STES system	Storage medium	Geological requirements	Storage capacity (kWh/m³)	Advantages	Disadvantages
Borehole	Rock, soil, sand, etc.	▲ Drilled underground ▲ Depth of 30-100m ▲ High heat capacity ▲ Favourable groundwater	15-30	▲ Used for both heating and cooling ▲ Less affected by atmospheric climate. ▲ Less surface area for vertical borehole. ▲ Feasible for Space heating of districts.	▲ Larger storage tank (3-5 times) compared to water tank ▲ High initial cost ▲ Suitable ground for drilling ▲ Typical performance obtained after 3-4 years.

5.4 Latent Heat Storage

The latent heat energy storage system basically depends on the principle of latent heat of fusion during the phase change process. In the phase change process, the latent heat is absorbed by phase change materials during melting, and on solidification, the stored heat is released. Therefore, the high heat of fusion plays the most important role in latent heat storage systems. The phase change of materials can be achieved through the following routes like solid-solid, solid-liquid, liquid-gas and solid-gas. The solid-liquid phase transition is the most suitable one considering latent heat as well as the small volume change during phase transition. The desired properties for a phase change material are high thermal conductivity, melting temperature range suitable for application range, congruent melting, no sub cooling, highly stable, non-toxic and non-corrosive.

Latent energy heat storage has been gaining much attention because it can provide high energy storage density per unit mass or per unit volume. The latent heat storage system is preferred over sensible storage due to the attainment of nearly isothermal condition over the charging and discharging process.

The energy store by latent energy source is given by Eq. (5.5),

$$Q = m \times L \qquad (5.5)$$

Where m = mass in kg and L is the specific latent heat (KJ/kg)

The storage capacity of PCM can also be formulated as Eq. (5.6) and (5.7) [1].

$$Q = \int_{T_i}^{T_m} mC_p \, dT + ma_m \Delta h_m + \int_{T_m}^{T_f} mC_p \, dT \qquad (5.6)$$

$$Q = m[C_{sp}(T_m - T_i) + a_m \Delta h_m + C_{lp}(T_f - T_m)] \qquad (5.7)$$

where, Q = stored energy(J),
m = mass of heat storage medium (kg)
C_p = Specific heat (J/kgK)
T_i = Initial temperature (°C)
T_m = Melting temperature (°C)
T_f = Final temperature (°C)

a_m = fraction melted
Δh_m = heat of fusion (J/kg)
C_{sp} = average specific heat between T_i and T_m
C_{lp} = Average specific heat between T_m and T_f

Organic materials and their eutectic mixtures are a more preferred option for TES as most of the organic materials melt in the range of useful purpose of human comfort, chemically stable, non-toxic, non-corrosive and readily available. Organic PCM can be broadly classified in two ways: Paraffin and non-Paraffin. Paraffin is the most used organic materials. Paraffin is a saturated n—alkane aliphatic hydrocarbon (CH_3-(CH_2)(n-2)-CH_3), where n is the number of carbon atoms. In practical applications, technical grade paraffin wax by the oil refinement industry are used as TES material [18]. The non-Paraffin category is the most used material in the study of heat storage. This category includes fatty acid, esters, alcohols and glycols and each material possesses distinctive properties of its own. The flammability nature of non-paraffin materials is a major drawback for using them in practical applications. Salt hydrates and metallic are the majorly used non-organic types of PCM. An eutectic mixture of PCM is the composition of two or more substances that possess a lower melting point than that of the individual PCM.

PCM storage systems have been used in both active and passive ways to meet the energy storage demand for different applications. PCM integrated with solar water heating system, solar air heating system, solar greenhouse are some of the well-known active modes of applications. The passive mode is mostly seen in the design of buildings like Trombe wall, PCM wallboards, PCM shutter, bricks incorporated with PCM, and ceiling boards, etc.

5.4.1 Enhancement Methods for Thermal Conductivity Enhancement

Poor heat transfer is one of the challenges for PCM as it affects the charging and discharging process. Several technologies have been employed to rectify this issue of PCM. Some of these technologies include microencapsulation and macroencapsulation, fins, heat pipes and porous matrices like metal foam. To increase the thermal conductivity of solid-phase PCM various fillers have been applied, namely, carbon-based fillers, metal-based fillers and ceramic fillers. Attempts to solve the problem of leakage control of PCM during melting have been made with methods like macro and microencapsulation, form-stable PCM with polymer matrix blending technique and form stable PCM chemical grafting technique [19]. Due to the porous nature of biochar, it can also be an option for the manufacturing of form-stable composite material. Composite phase change materials have been developed to overcome the low thermal conductivity issue in organic PCM. Graphite and nanocomposites are the most used composite PCM where the particles are dispersed in PCM. Form stable composite is another area of preparing composite where the stability of PCM is ensured by dispersing PCM into the porous matrix of some carbonaceous material.

5.4.1.1 Macro and Microencapsulation

Microencapsulation is done either by coating the droplets of PCM or by embedding the PCM in a homogeneous or heterogeneous matrix. Microencapsulation provides a physical barrier to prevent leakage from the core material as shown in Figure 5.6. Different shapes of particles can be obtained through different techniques starting from a simple sphere to irregular matrix. Chemical methods are more suitable for producing microencapsulated PCM due to its properties. Microencapsulation of PCM provides PCM thermal and mechanical stability. Coacervation, suspension polymerization, emulsion polymerization, polycondensation and polyaddition are some of the known methods of applying microencapsulation. Different methods available for producing microencapsulated PCM (MEPCM) have been mentioned in Table 5.3.

MEPCMs are used as building materials for energy conservation purposes due to their better heat transfer abilities and no leakage during phase transition. The increase in the surface area provided by microencapsulation is the prime reason for a better heat transfer rate than ordinary PCM. PCM slurry is another noble advancement where the normal working fluid is replaced by PCM slurries in many cooling and heating applications. Along with the properties of a working fluid, PCM slurry gives the advantage of latent heat storage in the working temperature range.

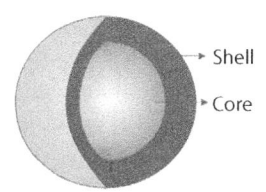

Figure 5.6 Encapsulated PCM [3].

Table 5.3 Method available for producing MEPCM [2].

Methods available	Different types	Characteristics
Physical	Spray drying Centrifugal processes Fluidized bed processes	Can produce microcapsules up to 100 μm.
Chemical	*In situ* polymerization Interfacial polymerization Suspension like polymerization Complex coacervation others	Size range from 5 and 100 μm Best quality in terms of diffusion tightness of walls. Microcapsules mostly in the range if 20-30 μm.

Similar to microencapsulation, macrocapsulation is the process of encapsulation of PCM in larger dimensions (more than 5 mm) capsules like spherical balls, rectangular sheets, etc. Macroencapsulation ensures high strength and durability over microencapsulation. It finds its wider applicability in building applications owing to the fact that it can be shaped in desired shapes in the form of pouches. The containers of encapsulation are selected based on the requirement, but the containers must possess properties like high thermal conductivity, durable and flexible, non-corrosive and fire resistance. Macroencapsulation is done through containment methods where PCM is filled in the required shape of containers. The main concern with this type is that the change of volume during the expansion process in a phase transition. This type of PCM finds applicability in buildings in increasing the latent heat storage capacity and thermal regulation [3].

5.4.1.2 *Addition of Fins*

The use of extended surfaces or fins to increase the heat transfer rate is a well-known technology. Fins provide additional surface area for heat transfer, which enhances the natural heat transfer rate. The design of the fin effects not only the convention but also conduction heat transfer. The fins can be applied to both the internal and external surfaces of the tubes of carrying heat transfer fluid. Length, spacing and thickness of fins are some of the critical parameters to be fixed before implementation [4]. Some studies confirmed that longitudinal fins and thinner fins are most effective [5].

5.4.1.3 Multiple PCM Technology

Multiple PCM technology involves the arrangement of PCM with different melting temperature in a systematic order. The main driving cause of the heat flow between PCM and fluid is the temperature difference between the two of them. In the case of a single PCM, the temperature difference gets decreased with the flow direction, which negatively affects the heat transfer rate. The modification is done by arranging the multiple PCM in decreasing order of their melting points, respectively. Such an arrangement provides nearly constant temperature difference throughout the flow direction. During the solidification process, the flow direction is made opposite to that of the melting phase. Constant heat flux is achieved with the combination of multiple PCM. The factors to be considered in such arrangements are the number of PCMs used and the appropriate selection of melting points of the PCM. An example of such arrangements includes rectangular airflow channels separated by PCM slabs for free cooling in air-conditioning systems [6]. Besides getting a better energy transfer rate, multiple PCM reduces the temperature fluctuations of the existing fluid [7].

5.4.1.4 Immersion Through Material Pores

The composite formation is a technology to enhance the properties mainly to rectify the heat transfer problem. Composite formation happens mostly by the methods of direct impregnation, vacuum impregnation and chemical solution intercalation. The composite formation is the formation of new material by combining PCM and porous material (metal foams, expanded graphite) by physical bonding. The formed composite possesses the same chemical properties in terms of crystallinity, functional groups as the parent PCM. The process of preparing involves the melting of the PCM in the matrix of porous materials so that the melted PCM can get trapped in the pores. Vacuum impregnation is the most used method as it ensures the maximum probability of opening the pores. Besides, enhancement of the thermal conductivity of the PCM composite formation also prevents leakage during the melting phase. Biochar as a supporting porous matrix has been used recently by researchers [8]. The thermal conductivity of PCM is mainly dependent on the matrix of the PCM and collective phonon motion over a long distance in a perfect crystal. The absence of proper alignment of molecules and interaction of atoms in the PCM matrix during the phase change process results in slower motion of phonon, and thus leads to the lower thermal conductivity of pure PCM [9]. The reason behind this enhancement can be explained in two ways. The first reason is the addition of foreign particle can make an orderly atomic structure, which promotes phonon motion in a particular direction over long distance [10]. The second reason is the filling of PCM material into the pores of porous materials. Thus the resistance offered by air pores has been reduced, and the thermal conductivity increases. The addition of metal powder can further increase the thermal conductivity. This is because of the high thermal conductivity range of metal powder. The metal powder acts as the bridge for thermal flow between adjacent biochar particles.

The melting temperature of the composite is found to be slightly lower than that of pure PCM. This is due to the reduction in the crystal nucleus with the addition of foreign particles. The heat of fusion is found to be normally the same as parent PCM. Composite PCM has good thermal stability. The presence of these surface tension and capillary forces not only assists in the shaping of the composite but also increases the thermal stability of the material. The composite formation by various researchers along with their specific properties are shown in Table 5.4.

5.5 Thermochemical Heat Storage

A Thermochemical heat storage system utilizes chemical energy as the storage medium, making use of redox chemical reactions. They are separated into chemical reactions and thermochemical sorption storage. Considering a redox reaction as shown in Eq. (5.8), in an endothermic reaction, reactant X is transformed into products Y and Z during the charging step. Compounds Y and Z are stored separately. Again due to discharging, Y, and Z recombine in a reversible process releasing energy. In the THS system, sensible heat storage is negligible compared to reaction heat, and there is no concern of heat losses. The sorption storage process includes adsorption and absorption processes. In the former process, gas bonds to solids surface without creating new material, while in the later process new compound is formed. Figure 5.7 shows different types of sorption thermochemical storage systems.

$$X + \Delta H_r \leftrightarrow Y + Z \qquad (5.8)$$

An example of such a process includes solid-gas reactions viz. decomposition of calcium carbonate, $CaCO_3$ (X) into solid oxide, CaO (Y) and CO_2 (Z); gas-gas reaction viz. ammonia dissociation reaction. In this process, NH3 gas (X) splits into gases N2 (Y) and H_2 (Z). These types of reactions

TES Systems for Cooling and Heating Applications   169

Table 5.4 Composite PCM formation and properties.

Author	PCM	Supporting matrix	Method employed	Thermal conductivity, latent heat and melting point of PCM	Thermal conductivity, latent heat and melting point of composite	Findings	Application of the composite recommend
Chen et al. [11]	Polyethylene glycol (PEG)	Almond shell biochar	Vacuum impregnation	0.251W/mK 189.06 J/g 56.93°C	0.402W/mK 59.37J/g 50.37°C	The 40% PEG content gives the best result	Energy storage material
Zhao et al. [12]	Polyethylene glycol (PEG)	Biological porous Carbon	Vacuum impregnation	0.427 W/mK 205.7 J/M 58.8°C	4.5 W/mK 158.8 J/M 56.5° C	Thermal properties are stable over 200 cycles	Building heat preservation material
Li et al. [13]	Polyethylene glycol (PEG)	ZSM-5	Vacuum impregnation	0.22 W/mK 192.6 J/g 59.5°C	0.54 W/mK 76.4J/g 56.3°C	The maximum content of PEG can reach 50% without leakage	TES
Wan et al. [14]	Palmitic acid	Pinecone char	Vacuum impregnation	0.2731 W/mK 84.74 KJ/kg 62.07°C	0.3926 W/mK 219.63 KJ/kg 59.25°C	PCM: Biochar ratio 6:4 is best in leakage test	TES
Zhang et al. [15]	Steraric Acid	Modified expanded vermiculite	Vacuum impregnation	0.21 W/mK 219.53 J/g 68.77°C	0.52 W/mK 134.31 J/g 67.12 °C	Mass ratio for composite is 10:1 (solid PCM: Supporting matrix)	Energy-efficient buildings

(Continued)

Table 5.4 Composite PCM formation and properties. (*Continued*)

Author	PCM	Supporting matrix	Method employed	Thermal conductivity, latent heat and melting point of PCM	Thermal conductivity, latent heat and melting point of composite	Findings	Application of the composite recommend
Zhang et al. [16]	Lauric-Stearic acid	Carbonized Corn cob	Vacuum impregnation	0.228 W/mK 199.6 J/g 37.5°C	0.441 W/mK 148.3 J/g 35.1°C	Maximum 77.9 wt% of PCM in carbonized corn cob without leakage	Energy efficiency in buildings
Song et al. [17]	Lauric Acid	Intercalated kaolinite	Solution intercalation	0.112 W/mK 161.3 J/g 44.3°C	0.101 W/mK 72.5 J/g 43.7°C	Maximum 48 wt % of lauric acid in matrix without leakage	
Karaipekli et al. [18]	Paraffin	Expanded perlite	Vacuum impregnation	0.05 W/Mk 275.9 J/g 36.18°C	0.15 W/mK 161.2 J/g 36.1°C	Addition of carbon nanotubes (1wt%) increased the thermal conductivity by 113.3%	Buildings and green house

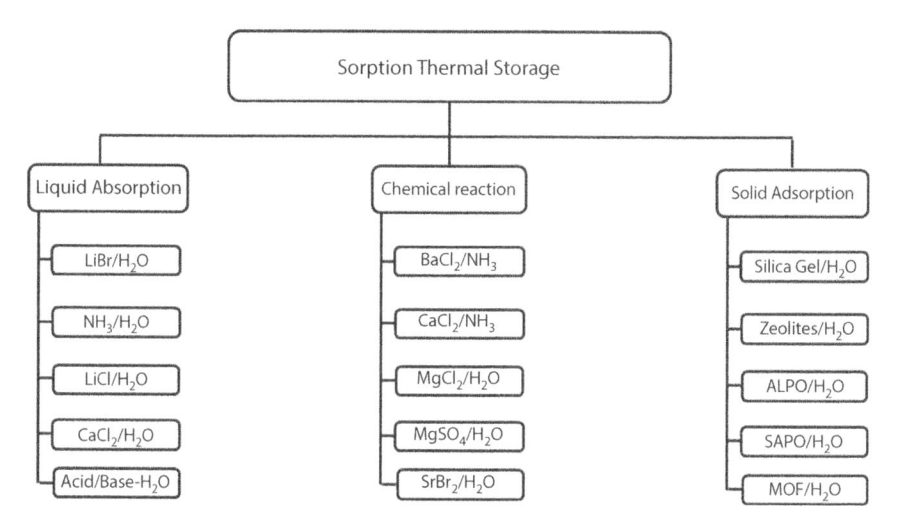

```
                    ┌─────────────────────────┐
                    │ Sorption Thermal Storage │
                    └─────────────────────────┘
         ┌───────────────────┼───────────────────┐
┌──────────────────┐ ┌──────────────────┐ ┌──────────────────┐
│ Liquid Absorption │ │ Chemical reaction │ │ Solid Adsorption │
└──────────────────┘ └──────────────────┘ └──────────────────┘
     │ LiBr/H₂O              │ BaCl₂/NH₃            │ Silica Gel/H₂O
     │ NH₃/H₂O               │ CaCl₂/NH₃            │ Zeolites/H₂O
     │ LiCl/H₂O              │ MgCl₂/H₂O            │ ALPO/H₂O
     │ CaCl₂/H₂O             │ MgSO₄/H₂O            │ SAPO/H₂O
     │ Acid/Base-H₂O         │ SrBr₂/H₂O            │ MOF/H₂O
```

Figure 5.7 Classification of sorption thermal storage system.

have very wide implications on the reactor design and system integration. For instance, the solid-gas reactions can be divided into open and closed systems. In an open system where the air is used as a reagent in redox, reactions are not stored separately but released into the atmosphere. On the other hand, carbonates, hydrides and hydroxides-based THS systems are included in a closed system as the reactant gases (CO_2, H_2 or H_2O) need to be stored in a separate tank. In this respect, the non-condensable gases require pressurized storage reservoirs compared to condensable gases. Open systems, when compared with the closed systems have the advantage of less complexity, and overall cost of facilities as gas storage can be avoided, but the partial pressure of oxygen has to be fixed at 0.21 atm, restraining redox reactivity.

THS, as compared to the SHS system, has 8-10 times higher storage density and two times higher as compared to the LHS system on a storage volume basis. However, for an efficient chemical reaction in the THS system, an efficient heat and mass transfer is required to and from the storage volume. As a result, this limits the overall storage volume of the THS system and is considered a key area for current research. THS system has a promising alternative of coupled high energy storage and low heat losses compared to SHS and LHS system. In addition, due to the superfluity of reversible reactions, the THS system provides higher flexibility in choosing the adequate system for a targeted temperature. However, due to complications of heterogeneous chemical reactions and somewhat higher handling of solids compared to liquid, THS systems are more complex.

Some of the properties which characterize the THS system are:

1. High reaction reversibility in both directions, without side reaction.
2. During storage, there no heat or entropy loss allowing long-term storage.
3. Greater flexibility in specific temperature ranges allowing a wider range of practical applications.
4. Stable kinetics and faster reaction rates during prolonged cycling.
5. The THS system requires heat to release sorbent (gas) from sorbent (matrix).
6. This system can be used either in summer for cooling (charging) or during winter for heating (discharging).

5.5.1 Absorption Cycle

Absorption is one of the sorption storage processes in which the adsorbate penetrates through the absorbent surface layer with a change of composition. The concentration of the storage density is closely linked with the concentration. Due to the suitability of the absorption cycle for low-grade heat utilization, absorption heat pumps are further developed compared to adsorption heat pumps. They belong to the indirect heat storage class where the storage heat utilizes both charging and discharging processes for converting heat into a different forms of energy, viz. mechanical, chemical. A schematic representation of the charging and discharging process in an absorption cycle is shown in Figure 5.8. Some of the commonly used sorbent/sorbate couples for adsorption systems are ammonia-water (NH_3-H_2O) and water-lithium bromide (H_2O-LiBr). Some of the other absorption couples, as suggested by Liu *et al.* include $LiCl/H_2O$, $Glycerin/H_2O$, $CaCl_2/H_2O$, KOH/H_2O, $NaOH/H_2O$. Among these $LiCl/H_2O$ is performed best in terms of efficiency and capacity but lower application in seasonal storage energy system due to its high price. On the other hand, $CaCl_2/H_2O$ is more cost-effective and has an appropriate regeneration temperature, but its storage capacity is very low.

The concept of salt in the porous matrix has been developed to improve the sorption properties of heat pumps. Tanashev *et al.* used a porous matrix to develop a composite inorganic salt where $CaCl_2$, LiBr and $MgCl_2$ were confined to pores of alumina and silica gel. Such sorbents, when used with water, are termed as Selective Water Sorbents (SWS), where one component is the host matrix (viz. alumina, silica gel, aerogel), and the

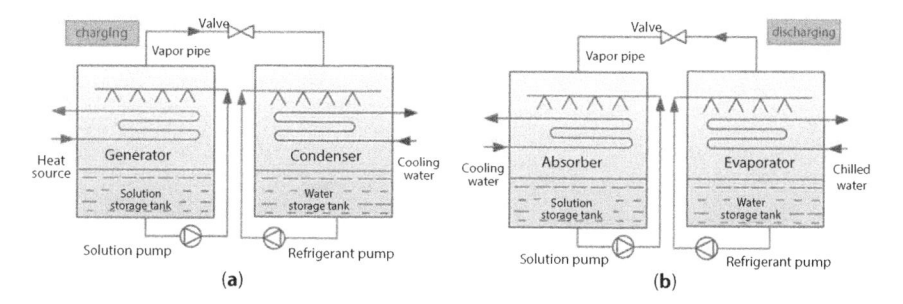

Figure 5.8 Schematic of absorption cycle (a) charging, (b) discharging.

other part is the inorganic salt placed inside the host matrix (viz. LiCL, $CaCl_2$, $MgCl_2$). The host matrix not only holds the adsorbent and prevents its dispersion, but some of them increase the surface area and increase the performance of salt/sorbate reaction, thereby enhancing the heat and mass transfer.

5.5.2 Adsorption Cycles

Adsorption is a thermochemical process in which liquid or gas is used as the adsorbate, which gets absorbed in the solid adsorbent (e.g., zeolites, silica gels, activated carbons). Adsorption heat storage can be applied for both direct and indirect storage methodologies. They belong to the indirect heat storage class where the storage heat utilizes both charging and discharging processes for converting heat into a different forms of energy, viz. mechanical, chemical. In this process, energy is stored in the form of adsorption potential energy without any loss until the adsorbate, and the adsorbent are kept separated.

Generally, two different configurations are developed for adsorption systems, viz. Open system and closed system. In a closed-cycle system, which occurs typically at low pressures, heat transfers take place from source to load through heat exchangers embedded within the adsorbent. During the charging process, the storage reactor where the adsorbate material is saturated with adsorbate is heated to release water vapour; the desorbed vapour is condensed and stored in a separate vessels. The condenser heat is then released to the surrounding or used to meet the load. The connection between the adsorber and condenser is closed once the charging is completed. On the other hand, to recover the stored energy discharging process is used where the liquid adsorbate reservoir now acts as the evaporator, and its connection with the adsorber is opened again. A low-temperature heat source provides the energy for evaporation of adsorbate,

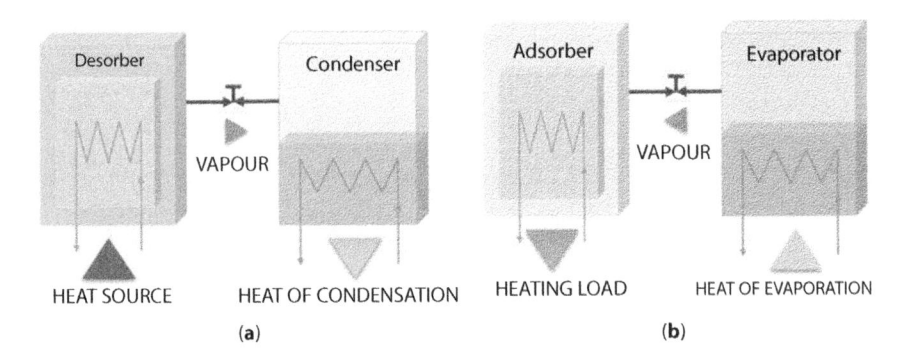

Figure 5.9 Schematic of a closed adsorption cycle (a) charging, (b) discharging.

which moves towards the adsorber and gets re-adsorbed by the adsorbent releasing energy, as shown in Figure 5.9.

In the open system, the heat exchangers are embedded external to that of the adsorbent. Similar charging/discharging process as that of a closed system with the exception the heat is provided by air flux through the adsorbent bed. During charging, the air, flux desorbs water vapour and exiting at lower temperature and humidity content. During discharging, the cold and moist air is provided to the adsorbent, which triggers the adsorption process and releases the stored energy from the adsorbent exiting as hot and dry air.

5.5.3 Chemical Reaction

Apart from the sorption process as discussed, another interesting method for THS is a chemical reaction. This process involves both reversible decomposition and synthesis reaction. Two of the commonly used candidates, $Ca(OH)_2$ and $Mg(OH)_2$ have high operating reaction temperatures (normally higher than 300°C) and are usually suitable for storing high-grade heat. Due to this higher temperature, no material is suitable for low-temperature solar application. However, with the addition of salt with a lower reaction temperature to pure $Mg(OH)_2$ viz. LiCl- modified $Mg(OH)_2$ as reported by Ryu *et al.* the temperature of the new mixture dropped by 70°C. Chemical reaction storages are still in the development stages and need further research in terms of high reaction enthalpy and high storage capacity of storage materials.

A comparative study of the three thermal systems is shown in Table 5.5.

Table 5.5 Comparison of three TES systems.

Parameters	Sensible storage	Latent storage	Chemical storage
Storage medium	➤ Water Tank, ➤ Aquifier, ➤ packed bed, boreholes	➤ Active ➤ passive	➤ Thermal-sorption (absorption and adsorption) systems ➤ Chemical reaction
Storage type	➤ Water, soil, gravel	➤ Organics and inorganics	➤ Metal hydrides, metal chlorides, metal oxides
Advantages	➤ Easy to control and relatively simple system. ➤ Reliable ➤ Environmental friendly	➤ Higher density compared to sensible TES system. ➤ Constant temperature thermal energy supply	➤ Energy density is highest ➤ Compact system ➤ Minimum heat loss.
Disadvantages	➤ Low energy density ➤ Problem of heat loss and self-discharge ➤ Geological requirements ➤ Site construction charges are higher	➤ Problem of crystallization ➤ Lack of thermal stability ➤ Higher storage material cost ➤ Problem of corrosion	➤ Under higher density condition problem of poor and heat mass transfer. ➤ Higher storage material cost. ➤ Uncertain cyclability

5.6 Application of Thermal Energy Storage Systems

Solar energy-assisted TES system transfers the stored heat to heat transfer fluid like air, water or some other designed fluid. The heat transfer fluid can be directly used for heating or cooling application or some heat exchanger is being provided.

5.6.1 Absorption Refrigeration System

This thermochemical TES system can effectively be utilized for meeting the heating and cooling demands in buildings. The system consists of a reversible absorption system that can be used in cooling mode as a single-effect absorption cycle and in heating mode as a heat transformer. All the components are both the systems that are the same except the direction of flow inside the system changes. Most of the absorption system utilizes flat plate or solar collectors as heat sources. LiBr-H_2O, where LiBr is used as absorbent and H_2O as absorbate is more suitable for low-cost solar application because of its lower operating cost. The absorption system is similar to conventional vapour compression but differs in the pressurization phase. Instead of a mechanical compressor, as in the case of vapour compression system, the absorption system completes the pressurization by dissolving the refrigerant in the absorbent. In LiBr-H_2O system, the solution used is very concentrated as the solubility limit is quite high. As shown in Figure 5.10, a solar collector is used as the heat input system to the generator of the vapour absorption refrigeration system (VARS). The system process involves an absorber, generator, condenser, the evaporator as the primary components.

The absorber contains the LiBr-H_2O mixture, which is delivered by a liquid pump to the generator. The solution in the absorber is a strong solution because of a higher percentage of refrigerant and is passed through a heat exchanger where its temperature increases. The absorbent-refrigerant mixture is heated in the generator from the heat source of the solar flat plate collector and it converts water into vapour. The absorbate viz. LiBr is left-back and is sent back to the absorber through a heat exchanger. The water vapour from the generator is then condensed in the condenser, and the heat is rejected, whereby the liquid condensate through an expansion valve is directed to the evaporator. In the evaporator, the heat for the load (cooled space) is absorbed and evaporates the refrigerant at low pressure. This creates a cooling effect. The vaporized refrigerant is then sent back to the absorber, where it again mixes with the weak LiBr, and thus, the process continues. Here the cooling effect achieved in the evaporator is

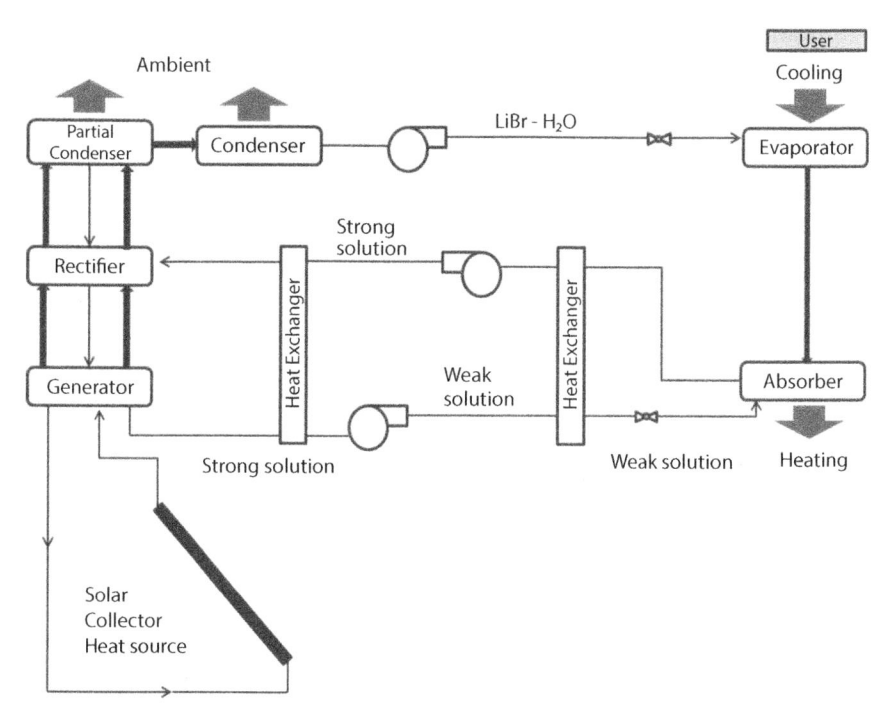

Figure 5.10 Working of a conventional VAR system with LiBr/H$_2$O as working fluid.

applied for either chilling air flow or water flow and further to deliver it for space cooling in buildings.

5.6.2 Solar Pumps Application in Space Cooling/Heating

Solar pumps find prominent application in space heating as well as cooling when used along with sensible TES systems, particularly Aquifer SHS systems and water-based systems. This section describes the working heat pump when used along with the Aquifer TES system for winter heating and summer cooling. The system consists of a typical aquifer arrangement coupled with a heat pump and cooling tower, as shown in Figure 5.11. Warm water is withdrawn from the low-temperature reservoir, i.e., the hot well at a particular temperature. As per the requirement of heat load for space heating, a portion of the warm water is pumped through the heat pump and transfers the heat to the building. The remaining portion is sent to the cooling tower, where it is cooled to a lower temperature. The cooler water from the heat pump along with the water from the cooling tower is sent back to the cold well for summer cooling operation. As a type of system has

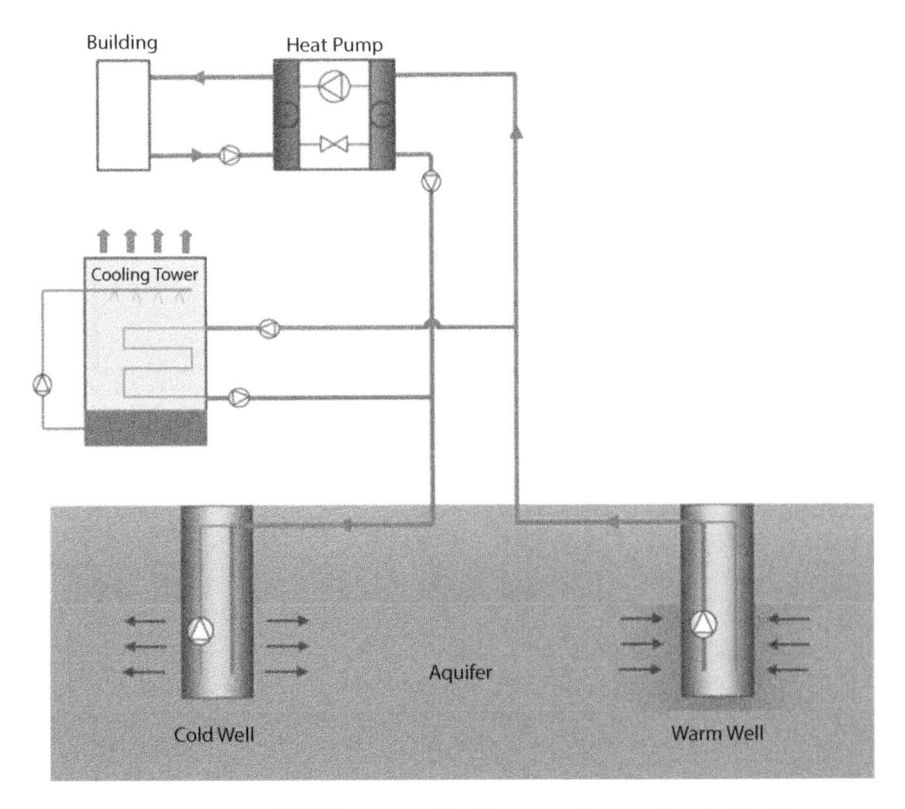

Figure 5.11 Integration of ATES system with solar pump for space heating/cooling.

been developed by Ghaebi *et al.* in Tehran. Paksoy *et al.* reported that such a combined heat pump system could result in an increase in COP by 60%.

5.6.3 Solar Pond Integrated Packed-Bed TES System for Space Heating

A solar pond is a large solar energy collector that stores solar energy through salt gradient technology. This stable salinity gradient prevents heat loss by thermal convection and traps the solar gradient. Solar pond consists of four different zones; the upper convective zone (UCZ), which acts an as solar collector with low salinity; the non-convective zone (NCZ), which occupies most of the depth of the pond; the lower convective zone (LCZ), which acts as highest salinity and the Ground zone (GZ). The NCZ acts as insulation between the LCZ and comparatively cooler UCZ. Along with the depth of the pond, the salt concentration increases, and once the pond is established, the solar radiation penetrates the zone and gets trapped in

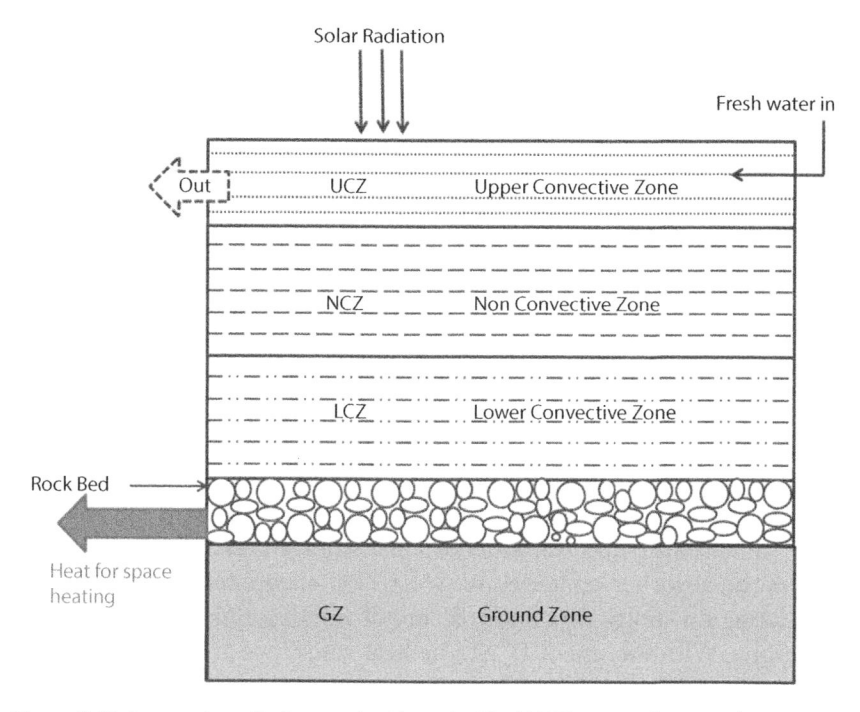

Figure 5.12 Integration of solar pond with packed bed THS system for space heating.

LCZ. In order to supply energy during winter seasons, the thermal storage needs to be efficient, and hence with the incorporation of a packed bed in the LCZ, the thermal energy generated can be stored for space heating application. The integrated system is as shown in Figure 5.12. In the case of rod bed being used as a heat storage medium, the effect of bed material, its porosity and bed thickness, plays a significant role in enhancing the thermal energy storage. Juwayhel and Refaee reported the Bakelite bed material of thickness 30% of that LCZ and porosity of 30% results is the enhancement of storage temperature by 22%.

5.6.4 Solar FPC

Solar FPC is a promising technology for the collection of solar energy. It is the most widely used and fundamental in the field of solar thermal energy harvesting systems for heating of water. Solar FPC basic design consists of a black absorber plate, a transparent glazing glass, tubes under the absorber plate, proper side and bottom insulation and the supporting structure. Solar radiation coming through the glazed surface is absorbed by the black surface and transfers to the working fluid inside the tubes. Therefore, the tubes are attached to the plate in such a manner that minimum heat transfer resistance is offered. The absorber is mostly made up of highly coating and highly conductive materials. The glazing surface is transparent to the short wave radiation of sunlight and opaque to the long wave re-radiated radiation wave energy. Though water is used mainly as the heat transfer fluid, the use of some antifreeze mixture, air and phase change liquids have also been investigated [19].

The integration of TESS with FPC can provide the economic feasibility of the system for the application of low-temperature water heating or space heating systems (Figure 5.13). Sensible energy storage needs a large volume of storage tanks while the PCM integration does require a compact design [20]. There are three types of design available for FPC with PCM, i.e., under the absorber plate, with the flow line or a separate storage unit. Integrated collector storage is a technology where solar FPC is connected with the storage tank itself [21]. Direct integration of PCM on FPC not only eliminates the need for an additional storage tank for PCM but also reduces the absorber temperature. Solar FPC encounters the problem of overheating on sunny days and freezing of working fluid in extreme cold conditions. With the use of PCM, the heat resistance and frost resistance can be accomplished [22].

The available solar cooling technology with FPC is mainly categorized into three types, namely, absorption chillers, adsorption chillers and desiccant cooling systems. The same system for space heating in winter can be used as solar Cooling in summer. High-temperature liquid FPC is the source of providing regeneration to the working refrigerant to the absorption chiller system. The refrigerant works on the same principle as a simple refrigerator but without the compressor. Adsorption chillers use

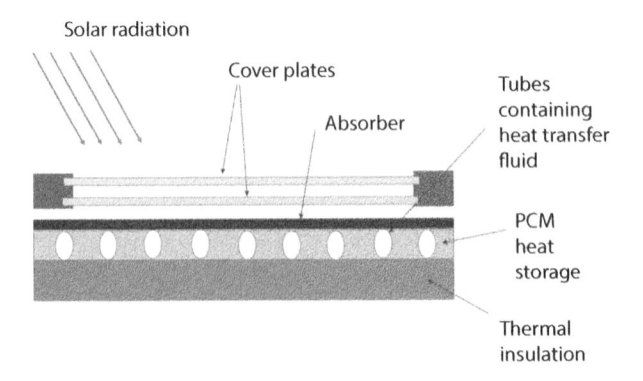

Figure 5.13 Solar FPC with LTES.

solid refrigerant instead of liquid refrigerant in absorption chillers. In the region where summer cooling is required and humidity are very high; then the solar desiccant system is a preferable option. When moist air passed through the desiccant system, the moisture is absorbed, and the air is cooled to the level where evaporative cooling is possible [21].

5.6.5 Solar PV/T

PV/T system is one of the methods, as shown in Figure 5.14, where the electrical efficiency of the PV cell is improved by cooling the PV cells and using the waste heat for some other useful purposes. Heat transfer fluid can either be air or water, which are suitable for applications like solar heating, water desalination, solar greenhouse, solar still and building heating system, etc. [24]. The PV/T technology is found to be more cost-effective than separate solar electricity and thermal generator. Different PV/T technology and their classification is shown in Figure 5.15.

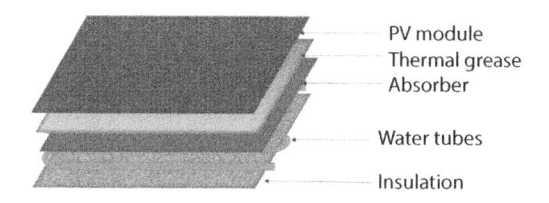

Figure 5.14 The components of Solar PV/T (water-based) [23].

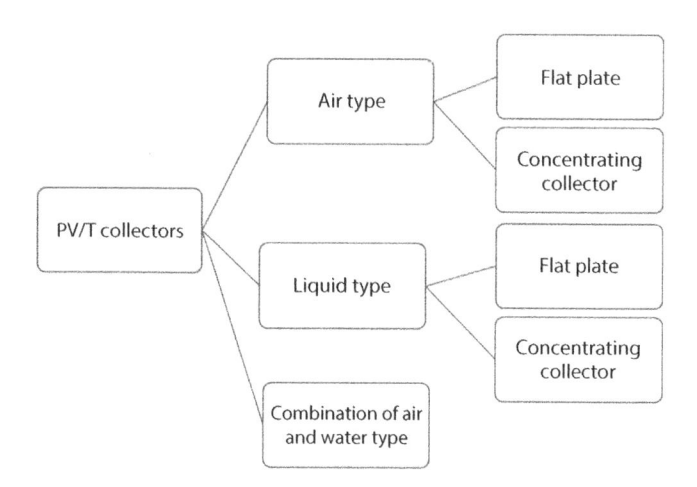

Figure 5.15 Classification of PV/T collectors [25].

Limited research can be found out to study the TESS integrated with PV/T for storage. The use of PCM and different nano-based composite PCM is the recently gaining popularity due to high energy density and high thermal and electrical efficiency of the system. A few findings on TES-based PV/T are shown in Table 5.6.

To minimize the electricity consumption in air-conditioning systems, a solar PV/T system with water or air as a heat transfer medium is a viable solution. PV/T technology can successfully eliminate the need for electrical energy to maintain the cooling load of buildings when attached to building façade or rooftops. Also, the use of water as a cooling fluid of the PV/T module is also suitable as a space heating option through a heat exchanger. The use of solar PV/T can reduce the high energy input for a compressor in the conventional cooling by the air-conditioning system. The heat gain from the PV/T module is utilized to evaporate the refrigerant of the cooling system. Solar PV/T system is useful for the desiccant cooling applications.

Table 5.6 TES based PV/T in building applications.

References	Type of thermal energy storage (TES) used	Important findings
Aldubyan et al. [9]	Borehole TES	Efficiency got improved by 4.1 and 4.7% in cold and hot climates, respectively.
Browne et al. [10]	PCM TES	The temperature of the water was 5.5 C higher with the use of PCM than the PV/T without PCM.
Lin et al. [11]	PCM TES (in ceiling ventilation system)	In the indoor thermal comfort of passive buildings without using air-conditioning systems with a maximum air temperature rise of 23.1 °C from the PVT collectors.
Lin et al. [12]	PCM (in building walls)	Coefficient of thermal performance enhancement increased to 43.4, 48.8 and 46.2% with an airflow rate of 2000 kg/h and a thickness of 20 mm PCM.

Some recent advancement in the field of PV/T technology is the circulation of refrigerating fluid through the PV/T module so that PV/T can act as the evaporator in the heat pump system. The evaporated vapour is then compressed in the compressor and passed through the condenser for space heating purposes. If the compressor runs on the electricity-driven from the PV module, then the whole heating system ensures sustainability. Microchannel based heat pipe PV/T system is another technology for heating where the PV/T module acts as the evaporation section. The other part of the pipe is the condensation section, where heat is released by the flowing fluid [26].

5.6.6 Solar Air Heater

The basic design of a solar air heater consists of a solar collector, inlet and outlet duct for entry and exit of air, proper thermal insulation for minimizing the loss of heat and a transparent glass cover. The whole assembly is fabricated mostly on sheet metal containers. The comparison of SAH with liquid FPC shows that as air is used as the working fluid so the system is simple for construction and operation. The corrosion and leakage problems are also eliminated in SAH compared to liquid FPC.

For continuous building heating operation and drying processes without any auxiliary heating systems, the need for a storage systems is inevitable [27]. An active, passive and hybrid types of SAH is mostly used. Both active and passive systems need storage systems [28]. Different types of latent and sensible heat storage systems have been developed by researchers in the recent past for enhancement of the efficiency of the solar air heater. Depending upon the construction, the system may operate on natural and forced convection. The blackened solar collector absorbs solar irradiation in the daytime, and heated inlet air is then passed through the exhaust system. The heated air can directly be used or can be circulated through a heat exchanger with secondary fluid for low-temperature applications like space heating, drying and some industrial processes. The intermittent nature of the sun prohibits the continuous supply of heated air, and therefore, the development of energy storage media plays an important role. The excess energy in the daytime is stored in the storage media, and on the nighttime or cloudy days, the air is circulated through the storage system. So the energy density of storage media is a critical factor for analysis. Solar air heating system with PCM is beneficial over water or pebble heat storage medium.

A hybrid thermal energy storage system is a technology where the PCM is charging both through solar and electrical energy simultaneously and able to reduce the electric consumption by 32% [29]. The study of the combined packed bed and latent heat storage has also been proposed where PCM in capsules is kept in insulated storage cylinders, and the fluid inlet temperature is proved to impact the heat transfer rate significantly [30]. The use of organic PCM like paraffin sometimes not preferable due to its poor thermal conductivity. Thus, some arrangements are provided on the SAH to enhance the heat transfer properties. The arrangements can be in the form of extended surfaces like fins. One modification involves the use of different shapes of absorber plates, for example, V corrugated absorber plate. The studies have proved that V corrugated absorber with PCM is able to provide hot air during the night at a temperature of more than 8-11°C of the ambient temperature [31]. Another such modification involves a flat plate collector with a staggered plate configuration [32]. Researchers also investigated PCM performance on double pass SAH where the lower duct is situated on the lower absorber surface, and backplate of heater for 1st pass of air and upper duct is placed between the glass cover and upper absorber surface for passing the recycled air [33].

5.7 Design Problems

Problem 1: A ten-story building is required to be space cooled during the summer season. A total of 50 tonnes of capacity cooling will be sufficient for complete refrigeration effect. Design an absorption refrigeration system using Ammonia-water as a refrigerant for space cooling of the building. If the system is coupled with a solar flat plate collector acting as a heat source, estimate the total energy required to be provided by the collector. The mixture value at various state points are given in Table 5.7 (1 tonne = 3.5 kJ/sec).

Solution
Capacity of refrigeration (Q_e) = 175 kW
For design of aqua ammonia refrigeration system assumptions considered,
Concentration of NH_3 in refrigerant, $X_r = 0.98$
Concentration of NH_3 in solution, $X_r = 0.42$
Concentration of NH_3 in absorbent, $X_r = 0.38$
Generator and condenser pressure, $P_H = 10.7$ bar
Evaporator and absorber pressure, $P_L = 4.7$ bar
Temperature of evaporator, $T_E = 2°C$
Temperature of condenser, $T_C = 54°C$
Temperature of absorber, $T_A = 52°C$
Temperature of Generator, $T_G = 120°C$

TES Systems for Cooling and Heating Applications 185

Table 5.7 Mixture values at different state points.

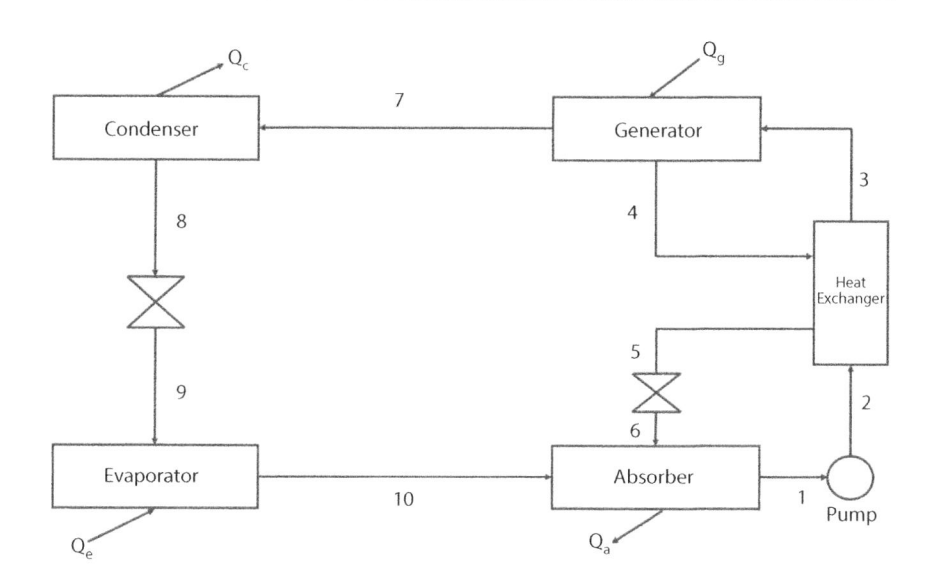

Temperature in °C	Pressure in bars	Specific enthalpy 'h' in KJ/kg
54	10.7	1135
54	10.7	200
2	4.7	200
2	4.7	1220
52	4.7	90
52	10.7	180
120	10.7	255
120	4.7	255

Calculations:
Point 8: (saturated liquid)

$$P_8 = 10.7 \, bar$$

$$X_8 = 0.98$$

$$T_8 = 54°C$$

$$h_8 = 200 KJ/kg$$

186 ENERGY STORAGE

Point 9:

$$h_8 = h_9 = 200 KJ/kg$$

$$T_9 = 2°C$$

$$P_9 = 4.7 bar$$

Point 10:
Saturation pressure in evaporator, $P_{10} = 4.7 bar$

$$h_{10} = 1220 KJ/kg$$

Heat extracted by evaporator,

$$Q_e = m_r(h_{10} - h_9)$$
$$\Rightarrow 175 = m_r(1200 - 200)$$
$$\Rightarrow m_r = 0.175 kg/sec$$

Mass balance equation,
Mass of solution (m_s) = mass of refrigerant (m_r) + mass of absorbent (m_w)

$$m_s = m_r + m_w \quad (5.5)$$

Now,

$$m_s X_s = m_r X_r + m_w X_w \quad (5.6)$$

Where, X_s, X_r and X_w are concentrations of solution, refrigerant and adsorbent respectively.

$$(m_r + m_w)X_s = m_r X_r + m_w X_w$$

$$\Rightarrow (0.175 + m_w)0.42 = (0.175 \times 0.98) + m_w(0.38)$$

$$\Rightarrow m_w = 2.45 \text{ kg/sec}$$

From eq. (5.1)

$$m_s = 0.175 + 2.45$$

$$\Rightarrow m_s = 2.625 \text{ kg/sec.}$$

Design of Evaporator
Let the inlet temperature to evaporator,

$$t_{h_1} = 30°C$$

At outlet temperature,

$$t_{h_2} = 5°C$$

$$\Delta\theta_1 = 30 - 2 = 28°C$$

$$\Delta\theta_2 = 5 - 2 = 3°C$$

$$LMTD = \theta_m = \frac{\Delta\theta_1 - \Delta\theta_2}{\ln\left(\dfrac{\Delta\theta_1}{\Delta\theta_2}\right)}$$

$$LMTD = 11.193°C$$

Heat extracted by evaporator from atmosphere,

$$Q_E = UA_E \theta_m,$$

where A_E is effective area of the evaporator
Assuming $U = 1000$ W/m², correction factor $F = 1$ and $Q_E = 175$kW

$$175 = 1 \times A_E \times 11.193$$

$$\Rightarrow A_E = 15.63 \text{m}^2$$

Let the number of evaporator tubes $(n) = 20$
Length of each tube $(L) = 1$ m
Effective area of evaporator, $A_E = n \times \Pi DL$

188 ENERGY STORAGE

Diameter of each tube, $\Rightarrow D = 25\text{cm}$
$(D) = 25\text{ cm}$

Design of condenser

At point 7, $P_7 = 10.7\,bar$
$X_r = 0.98$
$T_7 = 54°C$
$h_7 = 1135\,KJ/kg$

Heat rejected by condenser, $Q_C = m_r(h_7 - h_8)$
$Q_C = 163.6\text{ kW}$

Considering air as the cooling medium,
Let,
Inlet temperature of air is $(t_{C_1}) = 25°C$
Outlet temperature of air $(t_{C_2}) = 45°C$

$$\Delta\theta_1 = 54 - 25 = 29°C$$
$$\Delta\theta_2 = 54 - 45 = 9°C$$
$$LMTD = \theta_m = \frac{\Delta\theta_1 - \Delta\theta_2}{\ln\left(\frac{\Delta\theta_1}{\Delta\theta_2}\right)}$$
$$LMTD = 17.09°C$$

Heat rejected by condenser to atmosphere,

$$Q_C = UA_C\theta_m,$$

where A_C is effective area of the condenser
Assuming $U = 1000\text{ W/m}^2$ and correction factor $F = 1$

$$163.6 = 1 \times A_C \times 17.09$$

$$\Rightarrow A_C = 9.57\text{m}^2$$

Let the number of condenser tubes $(n) = 20$
Length of each tube $(L) = 1\text{ m}$
Effective area of evaporator, $A_C = n \times \prod DL$
Diameter of each tube $D = 15.24\text{cm}$

TES Systems for Cooling and Heating Applications 189

Design of Generator
At point 1, solution entering the pump,

$$P_1 = 10.7 \, bar$$

$$X_s = 0.42$$

$$T_1 = 42°C$$

$$h_1 = 90 \, KJ/kg$$

At point 3, High pressure saturated strong solution,

$$P_3 = 10.7 \, bar$$

$$X_s = 0.42$$

$$h_1 = h_3 \, 90 \, KJ/kg$$

At point 4, Weak solution leaves generator at saturation temperature,

$$P_4 = 10.7 \, bar$$

$$X_w = 0.38$$

$$T_4 = 120°C$$

$$h_4 = 225 \, KJ/kg$$

Energy balance for generator, $Q_G = m_r h_7 + m_w h_4 - m_s h_3$
$$\Rightarrow Q_G = 587 \, kW$$

Energy required to be supplied by the solar collector is 587 kW

Design of absorber,
Heat rejected in absorber,

$$Q_A = m_w h_6 + m_r h_{10} - m_s h_1$$

$$\Rightarrow Q_A = 602 \, kW$$

Problem 2. An apartment room of size (4 × 5 × 4m) is to be cooled by means of solar thermal driven chillers. The heat required for compressing the refrigerant is supplied by solar FPC through the PCM storage tank. Estimate the volume of the PCM storage tank if the room is facilitated with 2 windows (1×1m size each) on the north side. There is no other external heat gain from outside other than the window.

Solution.
Assumptions for the room and ambient conditions
Inside temperature of room = 27°C
Outside ambient temperature = 35°C
U_{wall} = 1.58 W/m²K
U_{roof} = 1.23 W/m²K
U_{foor} = 0.93 W/m²K
U_{glass} = 4.17 W/m²K
Solar heat gain of glass = 250 W/m²
Cooling load factor = 0.86
Internal shading coefficient = 0.87
We will neglect heat transfer through infiltration for our simplicity.
We will design the problem for 2 occupants in the room each having 70 W sensible heat/person
And 40 W latent heat/person.
Lighting load = 25 W/m²
Appliance load = 500 W (sensible) + 300 W (latent)

Sensible heat load calculation for the room
Heat transfer through the walls,

$$Q_{wall} = U_{wall} A_{wall} dT = 1.58 \times (4 \times 4 - 2 \times 1 \times 1) \times (35 - 27) = 176.96 W$$

Heat transfer through the roofs,

$$Q_{roof} = U_{roof} A_{roof} dT = 1.23 \times 5 \times 4 \times (35 - 27) = 196.8 W$$

Heat transfer through the floors,

$$Q_{fllor} = U_{fllor} A_{fllor} dT = 0.93 \times 5 \times 4 \times (35 - 27) = 148.8 W$$

Heat transfer through glass,

$$Q_{glass} = A_{glass}[U_{glass}(T_o - T_i) + SHG \times SC \times CLF] = 2 \times 1 \times 1 \times [4.17(35-27) + 250 \times 0.86 \times 0.87] = 440.82\,W$$

Sensible load due to occupants,

$$Q_{o,s} = occupantsno \times SH = 2 \times 70 = 140\,W$$

Sensible load due to lighting,

$$Q_{l,s} = A_{floor} \times lightingload = 5 \times 4 \times 25 = 500\,W$$

Total sensible heat load,

$$Q_{s,total} = 176.96 + 196.8 + 148.8 + 440.82 + 140 + 500 = 1603.38\,W$$

Total latent heat load,

$$Q_{l,total} = 40 \times 2 + 300 = 380\,W$$

Total head load,

$$Q_{total} = Q_{s,total} + Q_{l,total} = 1603.38 + 380 = 1983.38\,W$$

Considering a safety factor of 1.5,
Total cooling capacity needed for the room = 1983.38 × 1.5 = 2173.38W = 0.62 TR.

Let's consider the entire required energy is supplied by the PCM storage tank.

So, theoretically, the PCM storage tank should be the same capacity of the cooling requirement. But considering the possible heat losses in transmission, safety factor of 1.2 is provided.

Latent heat storage capacity = 1.2 × 1983.38 = 2380.05 W

PCM with a melting temperature of 25°C and a specific gravity of 0.9 is used as storage media. The latent heat for the chosen PCM is 230 KJ/kg.

Let V be the volume of the storage tank.
Now, 2380.05 = (0.9 × 1000 × V) × 0.230
V = 11.5 m³, Volume of the PCM tank.

Some of the recent advancement in space heating and cooling integrated with TES is mentioned in Table 5.8.

Table 5.8 Some applications of TES system by different researchers.

Researcher	Location	Storage type	Storage medium	Outcomes
Schmidt et al.	Hamburg, Germany	SHS	Water based	Solar collector of 3000 m² area with storage capacity of 4500 m³. The system was able to deliver heating effects to 14,800 m² living area—the tank material is made of concrete.
Oliveti et al.	Carabria, Italy	SHS	Water based	Size of solar collector 91.2 m² with a storage capacity of 500 m³. The system delivered heating effects to 1750 m² area. The tank material is made of reinforced concrete.
Colclough and Griffiths	Galway, Ireland	SHS	Water based	A small sized solar evacuated tube collector of 10.6 m² with tank capacity of 23.3 m³. The system provided heating effects for 215 m² living area. The water tank is underground.
Hahne	Stuttgart, Germany	SHS	Packed bed	Truncated cone-shaped pit geometry was considered. The solar collector of size 211 m² with storage volume 1050 m³ was utilized for providing both heating and cooling effects. Thermal insulation of pumice and plyurethane was used.

(Continued)

Table 5.8 Some applications of TES system by different researchers. (*Continued*)

Researcher	Location	Storage type	Storage medium	Outcomes
Mangold *et al.*	Steinfurt, Germany	SHS	Packed bed	The solar collector of 510 m^2 area and tank capacity if 1500 m^3 having truncated cone-shaped. Heating effects was provided to 3800 m^2 living area. Granulated foam glass in textile bags was used as insulating material.
Lundh and Dalenback	Anneber, Sweden	SHS	Borehole	The system provided heat supply to 50 residential units with the help of 2400 m^2 solar collector. Double U shaped pipes of 100 boreholes of 65 m depth with a total storage volume of 60000 m^3 were utilized.
Reuss *et al.*	Attenkirchen, Germany	SHS	Borehole	The system provided heat supply to 30 households with the help of 863 m^2 solar collector. Double U shaped loops of 90 boreholes of 30 m depth with a total storage volume of 10,500 m^3 were utilized.
Buchholz *et al.*	Berlin, Germany	TCHS	Absorption (MgCl$_2$/H$_2$O)	The system provided a significant storage density with economical material.
Bales *et al.*	Austria	TCHS	Adsorption (Silica/H$_2$O)	Long term feasibility of the system was lower due to low heat storage density.

(*Continued*)

Table 5.8 Some applications of TES system by different researchers. (*Continued*)

Researcher	Location	Storage type	Storage medium	Outcomes
Palomba et al.	Italy	TCHS	Adsorption (AQSOA FAM Z0$_2$/water)	System was utilized for providing both heating and cooling effects under different operating conditions. The storage density of the system reached 5 times that of water.
Fumey et al.	Switzerland	TCHS	Absorption (NaOH/H$_2$O)	Absorption occurred much slower than expected. Novel heat and mass exchangers were proposed for an effective heat storage system.
Harikrishnan et al. [34]	Chennai, India	Insulated TES tank with cylindrical encapsulation	Composite PCM (Paraffin and hybrid nanomaterials)	The heating and cooling rates increased with hybrid materials. For 1% mass of nanomaterials, the melting and freezing time was reduced by 29.8 and 27.7%, respectively.
Lavedeba et al. [35]	Latvia	PCM storage tank with cylindrical tubes	Paraffin	The average COP of the solar cooling system increased with PCM storage. Auxiliary heater using time reduced by 21%.
Vaidhyanathan et al. [36]	Northern region of India	PCM container	OM 37	The temperature of the room increased from 15°C to 23-24°C in the winter season with 20kg of PCM.

(*Continued*)

Table 5.8 Some applications of TES system by different researchers. (*Continued*)

Researcher	Location	Storage type	Storage medium	Outcomes
Helm *et al.* [37]	Munich, Germany	LHS	PCM calcium chloride hexahydrate ($CaCl_2 \cdot 6H_2O$)	The system replaces the wet tower of heat rejection in the absorption chiller in solar cooling and provides a dry cooling system by the capillary design heat exchanger.
Himpel *et al.* [38]	Garching, Germany	LHS	PCM calcium chloride hexahydrate ($CaCl_2 \cdot 6H_2O$)	Storage of two modules of the volume of 1.6m^3 storage capacity of 120 kWh was operated for heating and cooling for three years.

5.8 Conclusion

Considering the high energy demand in building heating and cooling applications worldwide, solar-based systems are a new boon to the scientific society. The intermittent nature of solar energy needs an efficient energy storage system for functioning in a continuous operation. SHS technologies in the form of water-based storage packed based storage, and aquifers are very widely used to store and use solar thermal energy. Solar pumps as a space heating option use aquifer SHS as energy storage media. Solar cooling achieved through vapour absorption system can replace the compressor work in a conventional cycle. Such type of cooling system is integrated with water-based storage or PCM storage. Solar pond system is mostly used with rock-based heat storage. As the energy storage density of PCM or LHS is more than the SHS and chemical storage, the use of PCM results in a compact and simple design. The low thermal conductivity is the main concern for the practical application of PCM. The different enhancement methods involve composite formation, encapsulation, use of the extended surface, blending of multiple PCM with different properties and nano-enhanced PCM. Various numerical-based problems have been investigated for the practical application of PCM on building heating and cooling. Solar PV/T can be a better option than individual operation by a solar air heater and solar FPC. Solar PV/T with PCM storage technology is a way forward technology as it can provide efficient electrical energy output and thermal performance. There is an excellent scope of combining both SHS and LHS to meet the energy storage demand for buildings heating and cooling. Recent advancements proved that absorption and adsorption types of solar-driven thermal chillers with the integration of both SHS and LHS have the potential to provide the required energy demand for building cooling.

References

1. J.A. Duffie, W.A. Beckman, *Solar Engineering of Thermal Processes*, 4th Edition, 2013. http://eu.wiley.com/WileyCDA/WileyTitle/product Cd-0470873663.html.
2. W. Su, J. Darkwa, G. Kokogiannakis, Review of solid – liquid phase change materials and their encapsulation technologies, *Renew. Sustain. Energy Rev.* 48 (2015) 373–391. https://doi.org/10.1016/j.rser.2015.04.044.
3. P. Kumar, S. Rathore, SK. Shukla, Potential of macroencapsulated PCM for thermal energy storage in buildings: A comprehensive review,

Constr. Build. Mater. 225 (2019) 723–744. https://doi.org/10.1016/j.conbuildmat.2019.07.221.
4. R. Senthil, M. Cheralathan, Natural heat transfer enhancement methods in phase change material based thermal energy storage, 9 (2016) 563–570.
5. M. Al-maghalseh, K. Mahkamov, Methods of heat transfer intensification in PCM thermal storage systems: Review paper, 92 (2018) 62–94. https://doi.org/10.1016/j.rser.2018.04.064.
6. A.H. Mosaffa, C.A.I. Ferreira, F. Talati, M.A. Rosen, Thermal performance of a multiple PCM thermal storage unit for free cooling, *Energy Convers. Manag.* 67 (2013) 1–7. https://doi.org/10.1016/j.enconman.2012.10.018.
7. P. Sivasamy, A. Devaraju, S. Harikrishnan, Science Direct Review on Heat Transfer Enhancement of Phase Change Materials (PCMs), *Mater. Today Proc.* 5 (2018) 14423–14431. https://doi.org/10.1016/j.matpr.2018.03.028.
8. D. Das, U. Bordoloi, H. Hihu, P. Kalita, A novel form stable PCM based bio composite material for solar thermal energy storage applications, *J. Energy Storage.* 30 (2020) 101403. https://doi.org/10.1016/j.est.2020.101403.
9. B. Eanest Jebasingh, A. Valan Arasu, A comprehensive review on latent heat and thermal conductivity of nanoparticle dispersed phase change material for low-temperature applications, *Energy Storage Mater.* 24 (2019) 52–74. https://doi.org/10.1016/j.ensm.2019.07.031.
10. H. Babaei, P. Keblinski, J.M. Khodadadi, Thermal conductivity enhancement of paraffins by increasing the alignment of molecules through adding CNT/graphene, Int. *J. Heat Mass Transf.* 58 (2013) 209–216. https://doi.org/10.1016/j.ijheatmasstransfer.2012.11.013.
11. Y. Chen, Z. Cui, H. Ding, Y. Wan, Z. Tang, J. Gao, Cost-effective biochar produced from agricultural residues and its application for preparation of high performance form-stable phase change material via simple method, *Int. J. Mol. Sci.* 19 (2018). https://doi.org/10.3390/ijms19103055.
12. Y. Zhao, X. Min, Z. Huang, Y. Liu, X. Wu, M. Fang, Honeycomb-like structured biological porous carbon encapsulating PEG: A shape-stable phase change material with enhanced thermal conductivity for thermal energy storage, *Energy Build.* 158 (2018) 1049–1062. https://doi.org/10.1016/j.enbuild.2017.10.078.
13. C. Li, H. Yu, Y. Song, M. Zhao, Synthesis and characterization of PEG/ZSM-5 composite phase change materials for latent heat storage, *Renew. Energy.* 121 (2018) 45–52. https://doi.org/10.1016/j.renene.2017.12.089.
14. Y. chao Wan, Y. Chen, Z. xing Cui, H. Ding, S. feng Gao, Z. Han, J. kai Gao, A promising form-stable phase change material prepared using cost effective pinecone biochar as the matrix of palmitic acid for thermal energy storage, *Sci. Rep.* 9 (2019) 1–10. https://doi.org/10.1038/s41598-019-47877-z.
15. X. Zhang, Z. Yin, D. Meng, Z. Huang, R. Wen, Y. Huang, X. Min, Y. Liu, M. Fang, X. Wu, Shape-stabilized composite phase change materials with high thermal conductivity based on stearic acid and modified expanded

vermiculite, *Renew. Energy.* 112 (2017) 113–123. https://doi.org/10.1016/j.renene.2017.05.026.
16. W. Zhang, X. Zhang, X. Zhang, Z. Yin, Y. Liu, M. Fang, X. Wu, X. Min, Z. Huang, Lauric-stearic acid eutectic mixture/carbonized biomass waste corn cob composite phase change materials: Preparation and thermal characterization, *Thermochim. Acta.* 674 (2019) 21–27. https://doi.org/10.1016/j.tca.2019.01.022.
17. S. Song, L. Dong, Y. Zhang, S. Chen, Q. Li, Y. Guo, S. Deng, Lauric acid / intercalated kaolinite as form-stable phase change material for thermal energy storage, 76 (2014) 385–389. https://doi.org/10.1016/j.energy.2014.08.042.
18. A. Karaipekli, A. Biçer, A. Sarı, V. Veer, Thermal characteristics of expanded perlite / paraffin composite phase change material with enhanced thermal conductivity using carbon nanotubes, *Energy Convers. Manag.* 134 (2017) 373–381. https://doi.org/10.1016/j.enconman.2016.12.053.
19. R. Dobriyal, P. Negi, N. Sengar, D.B. Singh, A brief review on solar flat plate collector by incorporating the effect of nanofluid, *Mater. Today Proc.* 21 (2020) 1653–1658. https://doi.org/10.1016/j.matpr.2019.11.294.
20. Z. Chen, M. Gu, D. Peng, Heat transfer performance analysis of a solar flat-plate collector with an integrated metal foam porous structure filled with paraffin, *Appl. Therm. Eng.* 30 (2010) 1967–1973. https://doi.org/10.1016/j.applthermaleng.2010.04.031.
21. I. Renewable, E. Agency, *Solar Heating and Cooling for Residential Applications,* (2015).
22. D. Wang, H. Liu, Y. Liu, T. Xu, Y. Wang, H. Du, X. Wang, Solar Energy Materials and Solar Cells Frost and High-temperature resistance performance of a novel dual-phase change material flat plate solar collector, 201 (2019). https://doi.org/10.1016/j.solmat.2019.110086.
23. A. Farzanehnia, M. Sardarabadi, Exergy in Photovoltaic / Thermal Nanofluid-Based Collector Systems, (n.d.) 1–13.
24. W. Lin, Z. Ma, M.I. Sohel, P. Cooper, Development and evaluation of a ceiling ventilation system enhanced by solar photovoltaic thermal collectors and phase change materials, *Energy Convers. Manag.* 88 (2014) 218–230. https://doi.org/10.1016/j.enconman.2014.08.019.
25. M. Mumtaz, A. Khan, N.I. Ibrahim, I.M. Mahbubul, R. Saidur, F.A. Al-sulaiman, Evaluation of solar collector designs with integrated latent heat thermal energy storage: A review, 166 (2018) 334–350. https://doi.org/10.1016/j.solener.2018.03.014.
26. X. Zhang, X. Zhao, S. Smith, J. Xu, X. Yu, Review of R&D progress and practical application of the solar photovoltaic/thermal (PV/T) technologies, *Renew. Sustain. Energy Rev.* 16 (2012) 599–617. https://doi.org/10.1016/j.rser.2011.08.026.
27. P.K. Choudhury, Solar air heater for residential space heating, *Energy, Ecol. Environ.* 2 (2017) 387–403. https://doi.org/10.1007/s40974-017-0077-4.

28. V. V Tyagi, NL. Panwar, N.A. Rahim, R. Kothari, Review on solar air heating system with and without thermal energy storage system, *Renew. Sustain. Energy Rev.* 16 (2012) 2289–2303. https://doi.org/10.1016/j.rser.2011.12.005.
29. Z.A. Hammou, M. Lacroix, A new PCM storage system for managing simultaneously solar and electric energy, 38 (2006) 258–265. https://doi.org/10.1016/j.enbuild.2005.06.008.
30. R.. Nallusamy, Nallusamy & Sampath, Shobana & Velraj, Experimental investigation on a combined sensible and latent heat storage system integrated with constant/varying (solar) heat sources, *Renew. Energy.* 32 (2007) 1206–1227. https://doi.org/https://www.researchgate.net/deref/http%3A%2F%2Fdx.doi.org%2F10.1016%2Fj.renene.2006.04.015?_sg%5B0%5D=0V8BjMNXebU1eDkdRYqcai83MmsmbdCytYoti38Go7Xwzq8GiTMKyMI5Mfx6DC2kInlEbBnhVjDciahOwPL5n0A9kA.lYVNlFORH7L-zW4OojFkDaq9IroExHen8zUWmAuS6qtpUtyA2ewN6HYfnu0WTym9fOhpWlxp82enVeXP6tG7rg.
31. S.M. Shalaby, A.E. Kabeel, M.K. Elfakharany, Investigation and Improvement of Thermal Performance of a Solar Air Heater Using Extended Surfaces Through the Phase Change Material, 142 (2020) 1–6. https://doi.org/10.1115/1.4044565.
32. A. Reyes, Computational Simulation of the Thermal Performance of a Solar Air Heater Integrated With a Phase Change Material, 141 (2019) 1–9. https://doi.org/10.1115/1.4043549.
33. S.M. Salih, J.M. Jalil, S.E. Najim, Double-Pass Solar air Heater (DP-SAH) utilizing Latent Thermal Energy Storage (LTES) Double-Pass Solar air Heater (DP-SAH) utilizing Latent Thermal Energy Storage (LTES), (2019). https://doi.org/10.1088/1757-899X/518/3/032038.
34. SHKDS. Kalaiselvam, Thermal energy storage behavior of composite using hybrid nanomaterials as PCM for solar heating systems, (2014) 1563–1571. https://doi.org/10.1007/s10973-013-3472-x.
35. K. Lebedeva, L. Migla, Latent thermal energy storage for solar driven cooling systems, (2020) 1134–1139. https://doi.org/10.22616/ERDev.2020.19.TF273.
36. B.N. Vaidhyanathan A., Theoretical Modeling of Phase Change Material-Based Space Heating Using Solar Energy, *Adv. Energy Res.* 2 (2020) 6–24.
37. M. Helm, C. Keil, S. Hiebler, H. Mehling, C. Schweigler, Solar heating and cooling system with absorption chiller and low temperature latent heat storage : Energetic performance and operational experience. *Int. J. Refrig.* 32 (2009) 596–606. https://doi.org/10.1016/j.ijrefrig.2009.02.010.
38. M. Himpel, S. Hiebler, M. Helm, C. Schweigler, Long term test results from a Latent Heat Storage developed for a Solar Heating and Cooling System, (n.d.).

6

Optimistic Technological Approaches for Sustainable Energy Storage Devices/Materials

Benjamin Raj[1], Arya Das[1,2], Suddhasatwa Basu[1] and Mamata Mohapatra[1,2]*

[1]CSIR-Institute of Minerals and Materials Technology, Bhubaneswar, Odisha, India
[2]Academy of Scientific and Innovative Research (AcSIR), Ghaziabad, Uttar Pradesh, India

Abstract

The pursuit of a sustainable future can be realised by not only utilising available sustainable energy sources but also coupling them with sustainable energy technologies for final delivery. Batteries and supercapacitors have served as effective technologies for sustainable storage devices but still face fundamental and technological hurdles to be overcome. Batteries and supercapacitors are considered to be complementary devices which have often been coupled with each other for performance enhancement or used individually depending on the end application. Batteries and supercapacitors have recognized adequate market space with their rapid expansion of demand, but require intensive research for development of robust technologies to sustain the next-generation energy requirements. This chapter will focus on the current state of the art of energy storing devices on the basis of their market roll out and technological maturity along with highlighting the production and performance. The progresses and constraints in the development of technologies from material to device design will be elucidated. Finally the new emerging materials and technologies with adequate potential of future implementation as high-performing sustainable energy technologies will be discussed and evaluated.

Keywords: Sustainable energy, batteries, supercapacitors, electrochemical performances

---

*Corresponding author: mamata@immt.res.in

6.1 Introduction

Energy storage devices are a key desideratum for an efficient and affordable warehouse of sporadic renewable sources to mitigate the need for fossil fuels by rapid development of sustainable storage technologies. To meet the current and future demand for portable electric devices, electric vehicles and stationary energy storage for the electricity grid is driving developments in electrochemical energy-storage (EES) devices. Undeniably, several different types of energy storage devices such as batteries, capacitors including capacitors, batteries, flywheels, superconducting magnetic energy storage (SMES) systems and solid oxide fuel cells (SOCF) are currently in practice. However, technological innovation for achieving commercial affordability of such systems by modulating energy and power capabilities and long cycling lifespans is the requirement of the present era. Significant endeavors are reported to enhance the energy-storage capacity of EES with industrial intervention and to meet future demands in various emerging applications. Among all energy storage devices batteries and supercapacitors have drawn prodigious attention because of their high efficiency, excellent retention efficiency, ultrahigh energy density, power density and eco-friendliness for the development of clean and green energy technologies. LIBs and supercapacitors are highly suitable in portable electronic devices, hybrid electric vehicles and large-scale electric networks. Generally supercapacitors own higher power density than that of the conventional batteries whereas energy density of batteries outperform supercapacitors. The efficiency of the storage devices fully depends on the nature and surface morphology of the electrode materials. In this chapter we have briefly discussed knowledge about batteries and supercapacitors. The recent development of advanced materials towards the energy storage devices has been discussed and presented. This chapter also deals with the current market scenario to cater to the needs of next-generation energy applications.

6.2 Advancements in Supercapacitor Technology

The current ongoing R&D efforts focussing on a sustainable tomorrow highly rely on development of effective energy storage systems and their wide implementation. In pursuit of the same, Supercapacitors (SC) are expected to play a pivotal role and thus have been subjected to rigorous study [1–3]. Supercapacitors are energy storing devices endowing high power densities and extraordinary stability alongside power dissipation at very minute interval of time. The rapid power delivery enables their use in enormous applications demanding high current and specific power ranging from small portable electronics to medium power backups to large-scale modules for hybrid automobile systems [4–6]. The rapid storage and dissipation of charges in supercapacitors is on account of surface or near to surface charge storage mechanisms at the electrode surface. SCs can store charge electrostatically via (1) charge separation at electrode/electrolyte interface (EDLC) [7], (2) fast redox reactions near the surface of the electrode (pseudocapacitance) [8], or (3) hybrid by retaining both the mechanisms [9]. Carbon materials are usually employed as EDLC electrode materials and pseudocapacitive electrode include transition metal oxide, sulphides and conducting polymers but can also include various doped carbon materials [7, 10, 11]. Hybrid capacitance broadly means the combination of pseudo and double layer capacitive materials to nourish the advantages of both the storage mechanisms for improved overall capacitance [12]. One of the major characteristics of supercapacitor mechanism is the linear dependence of capacitance on the potential whilst batteries store all the charge at a given potential. This is to be sufficed by all materials for supercapacitive storage distinguishing them from battery storage mechanism [13, 14]. Based on the apt distinguishing of the storage mechanisms and contribution the conventional methods of assembling a supercapacitor have moved ahead to advanced assemblies to achieve enhanced performance and have been broadly categorized into (Figure 6.1):

   a. Symmetric Capacitors (identical electrode assembly – EDLC or Pseudo) [15]
   b. Asymmetric Capacitors (two unidentical electrode assembly – EDLC or Pseudo) [16]
   c. Supercapattery - (one supercapacitor electrode (EDLC or Pseudo) + one battery type electrode) [17]

The performance of SCs is greatly influenced by the electrode material, electrolyte and window of operation which in turn depends on the electrode and electrolyte compatibility and their conjugate operation range (Figure 6.2). Thus the advancement of novel supercapacitor technologies requires rational designing of electrode along with appropriate use of electrolyte to have significant impact on performance enhancement of SCs. Electrode materials for effective potential SC electrode demand high electrochemical active sites, porosity, high theoretical capacitance and good electrochemical conductivity [18–20]. A lot of research has focussed on

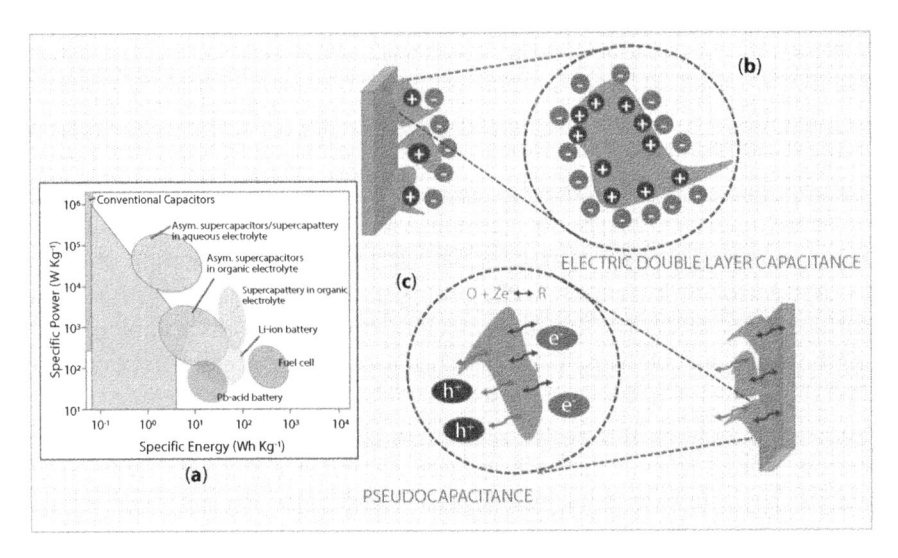

Figure 6.1 (a) Ragone plot of various electrochemical storing and conversion devices [17]. (b, c) illustrating the EDLC and pseudocapacitive mechanisms in supercapacitors, respectively.

tailoring the morphology and structure of diverse electrode materials for targeted storage mechanism and has displayed optimistic results [21–25]. The choice of electrolyte along with electrode selection is a major challenge for performance improvement of SCs. No electrolyte can be inferred as perfect; each has its own merits and demerits and its selection relies on electrode compatibility and end application. Aqueous electrolyte offers the maximum conductivity and capacitance but suffers from leakage issues and inferior energy densities. Organic electrolytes offer better energy densities with wide potential window but have lower ionic conductivity. Solid state electrolytes may resolve the leakage problem but still possess lower ionic conductivity. With extended research new electrolytes based on right proportion of water/co-solvent and salt have reflected promising capacitance, conductivity and wide potential stability [26–28]. There has been a rapid upsurge in the demand of supercapacitors providing the market transition to electrical vehicles and higher demand of portable electronics. Though the boom serves as a platform of greater opportunities for SCs there are a lot of challenges and constraints to address. This section provides an insight into the current global supercapacitor market, the challenges of commercialisation of technologies from lab to market and development of novel materials for possible commercialisation.

6.2.1 The Current Global Supercapacitor Market

Supercapacitors are receiving high market growth and the market share is expected to increase in future. This will be in spite of an expected decline in 2020 due to the global COVID-19 pandemic, after which the supercapacitor market is expected to soar. A report published by Lucintel group forecasts a global supercapacitor market of $3.5 billion by 2025 compared to $1.1 billion in 2019 with a Compound annual growth rate (CAGR) of 19 to 21% from 2019 to 2025, as shown in Figure 6.3 [29]. The majority of the market was dominated by pseudocapacitors in terms of revenue in 2019, and hybrid capacitors are predicted to contribute the most to the predicted CAGR of 21% till 2025. The noticeable factor that boosts the supercapacitor market development could be directly attributed to favourable government regulations and demand for sustainable energy systems. Moreover, there has also been rising demand for micro-supercapacitors and wearable supercapacitors, leading to estimates of lucrative prospects for supercapacitor market growth [30]. The electric double layer capacitor accounts for the largest share of the supercapacitor market with its wide use in automobiles and consumer electronics. The Asia-Pacific region has the highest demand for supercapacitors while Europe accounts for the largest production with the awareness of energy sustainable systems gradually raising the demand worldwide. This could be inferred from the recent advancements in processing and supply manufacturing illustrations. The Indian Space Research Organization (ISRO), established a centre for its technology to produce capacitors of variable capacitance values (2.5 V) to

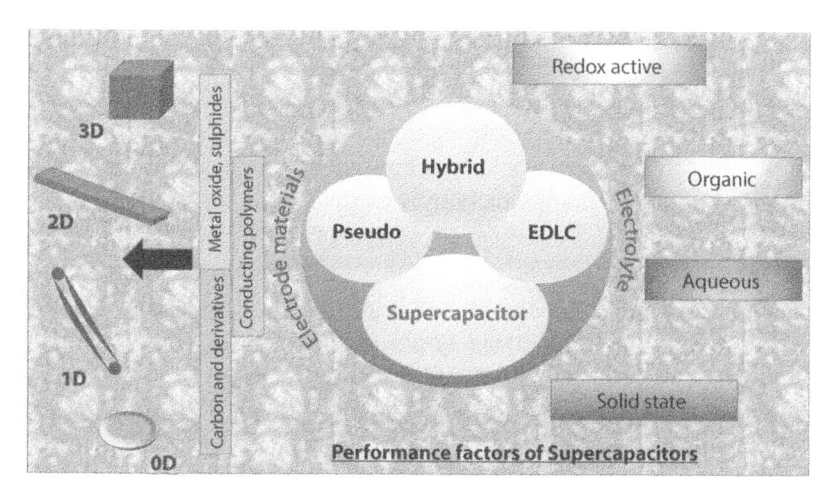

Figure 6.2 Factors affecting performance of supercapacitors.

206 ENERGY STORAGE

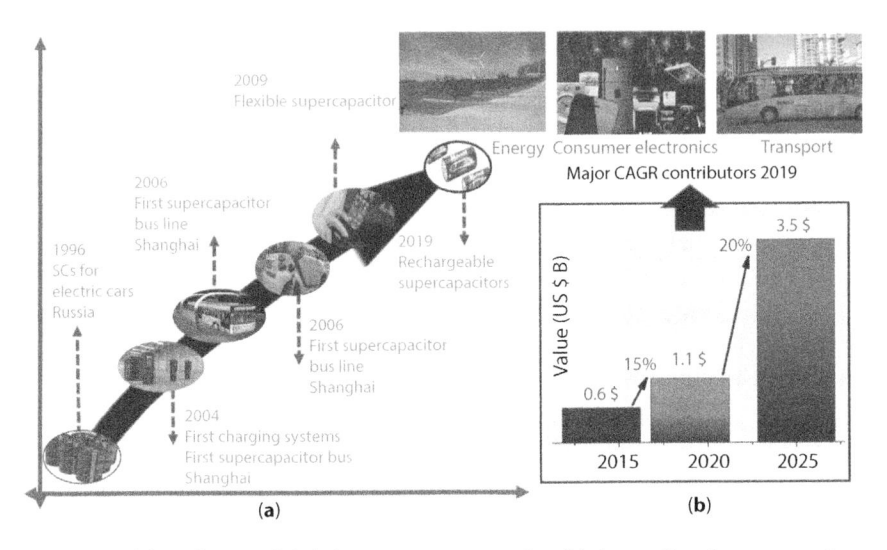

Figure 6.3 (a) Evolution of global supercapacitor market (b) the predicted supercapacitor growth with major CAGR contributors.

cater to its space applications. In another instance Tesla acquired the leading producer of supercapacitor in February 2019 to look for novel innovations for the evolving electric automobile industry [31].

As discussed in the earlier section, a supercapacitor consists of electrode and electrolyte, alongside separator and current collector to drive the current safely for final delivery. The majority of the available commercial supercapacitors are cylindrical with electrode materials loaded on aluminium foil rolled soaked in electrolyte and hard cased to prevent electrolyte evaporation and contamination. The loading, thickness of the active material and height of the electrode loaded roll is decisive in determining the end performance of the supercapacitor. In another assembly, pouch configuration can lead to higher power output by stacking electrode, separator and electrolyte of same polarity in parallel configuration into a plastic pouch. Although the manufacturing is complex and cost burdening compared to its cylindrical counterpart, it can produce considerable results. The two types of supercapacitor assembly are presented in Figure 6.4.

Commercial supercapacitors are predominantly produced by a handful of countries occupying the majority of the market share (Table 6.1). These products are categorised by product kind, module kind (10–25 V, 25–50 V, 50–100 V and >100V modules), their application. With advancement in the supercapacitor technologies >100V modules are expected to actively participate in the CAGR achievement. The commercial supercapacitors available from the major manufacturers are compiled and presented in

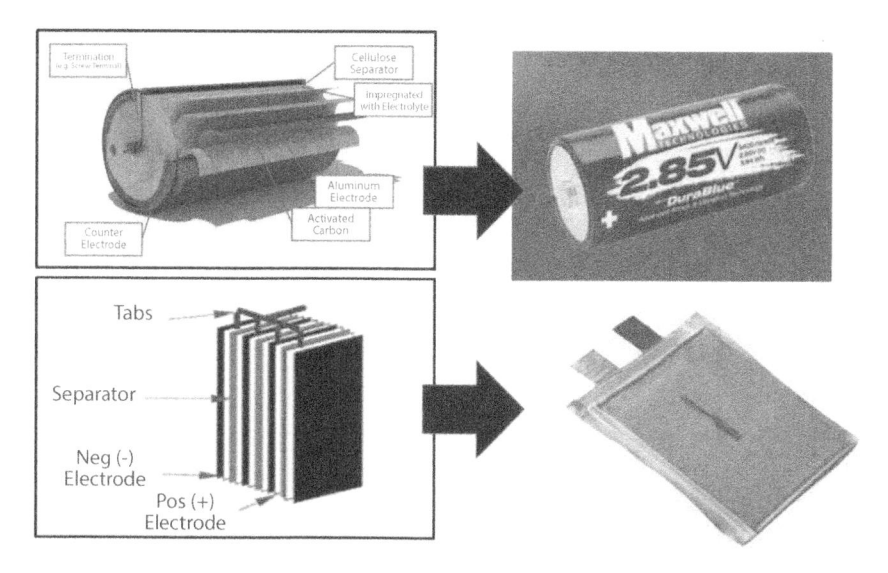

Figure 6.4 Two types of supercapacitor assembly representation: cylindrical and pouch type.

tabular form in Table 6.1. As evident from the table, the majority of the products are symmetric in nature with the highest voltage rating of 2.85V made available by Maxwell technologies. Energy density varies widely with a maximum of 7.5 Wh/Kg for LS Mtron cell with hybrid symmetrical cell form Yunasko offering high power ratings of 41 kW/Kg.

Though the discussion clearly indicates ample opportunities due to the exponential demand expectations of supercapacitors in coming years, the transition of lab-scale technologies to market faces enormous challenges to overcome and withstand as sustainable energy technologies among other competitors.

6.2.2 Challenges: From Lab to Market

Despite rigorous research exploring a wide range of materials and electrolytes bottlenecks still exist for successful transition of materials from lab to market even though they achieve energy and power densities comparable with or even better than existing energy technologies [40, 41]. The major challenges include:

    a. Most of the materials that are reported are not economically viable to be commercially employed. The major cost of a supercapacitor is the cost of materials, which can be

Table 6.1 Different major manufacturers and commercial capacitors available in the market.

Company	Country	Type	Device	C (F)	V (V)	Energy density	Power density
Maxwell [32]	U.S.A.	Sym.	Cell Module	1-3400 / 5.8-500	2.3-2.85 / 16-160	0.7-7.4 / 2.3-4	2.4-1.4 / 3.6-6.8
LS Mtron [33]	Rep. Korea	Sym.	Cell Module	100-3400 / 2.5-500	2.7-3 / 16-381	3.3-7.5 / 2.3-5	0.9-2.4 / 0.3-0.6
Nesscap Co. Ltd. [34]	Canada	Asym. Sym.	Cell Cell Module	50-300 / 3-5000 / 1.5-500	2.3 / 2.3-2.7 / 5-125	4.8-8.8 / 2-5.7 / 1-3.7	4.9-6.2 / 6-17 / 2.5-6.8
Panasonic Corp [35]	Japan	Sym.	Cell Coin	3.3-100 / 0.1-1.5	2.3-2.7 / 3.6-5.5	1.4-4.1 / 0.13-1.5	0.29-3.65 / 0.001
Vinatech [36]	Rep. Korea	Hybrid Sym.	Cell Cell Module	10-800 / 1-3000 / 2.5-60	2.3 / 2.5-3 / 16-144	4-12 / 0.7-6.5 / 1.7-5.6	0.4-0.8 / 1-11.3 / 0.4-4
Yunasko [37]	Ukraine	Hybrid Sym.	Cell Cell Module	1.3 Ah / 400-3000 / 13-500	2.7 / 2.7 / 16-90	37 / 4.7-6.2 / 3.9-2.1	4 / 7.1-41 / 3.3-11.3
IoxusInc [38]	U.S.A.	Sym.	Cell Module	1250-3150 / 21-500	2.7 / 16-162	4.5-6.3 / 2.2-3.8	23-34 / 0.2-0.4
SPS Cap [39]	China	Sym.	Cell Module	1-5000 / 0.5-500	2.5-2.7 / 16-2300	0.9-6.5 / 1.4-3.6	0.5-8 / 0.3-0.7

tackled by choosing earth-abundant cost-effective materials. Recycling and beneficiation of potential materials can also serve the purpose to a larger extent [42].
b. Complex synthesis methods of electrode materials is also a major drawback for commercial implementation. Thus simple production method with consistent quality is largely appreciated [43].
c. Benchmarking of electrode material is another major bottleneck with no standard performance assessment available. Thus it is advisable to adopt the known parameters of commercial supercapacitor such as mass loading of 10mg/cm^{2n} to actually know the performance in a larger picture of real applicability [44].
d. Lastly, all these factors should go through proper assessment including human resource cost, machinery cost, etc., to actually see the viability of employment in the present market. In case of extraordinary results regardless of the cost-effectiveness, employability is still possible given the market demand considering the high performance and market needs.

6.2.3 Current Trends and Opportunities

Several recently reported studies point towards development of exceedingly optimized and competent electrodes in terms of real applicability. Though it is difficult to ascertain whether these novel architectures could actually outperform their commercial counterparts and is ambiguous to compare, we put forward some of the recent materials with impending potential to be engaged as commercial supercapacitor electrodes. It is noteworthy to mention that there has been much work done in developing novel energy storage devices that would employ next-generation smart and intelligent technologies.

6.2.4 Composites and Novel Architectures

The rational tailoring and architectures can have significant impact on the performance of electrode materials and device performance in whole [45, 46]. Zhao *et al.* recently reported nickel cobalt sulphide – graphene hybrid (NCS-G-H) defect rich honeycomb like structures. The graphene incorporation not only served as a template for synthesis but also helped in the improvement overall conductivity and electrochemical active sites. The NCS-G-H reflected specific capacitance of 1186.7 F g^{-1} at 1 A g^{-1} which when assembled into a NCS/G-H//Activated carbon device delivers a 46.4 Wh kg^{-1} at a power density of 400 W kg^{-1} [47]. Hong et al. could grow binder free 1D single crystalline Cu$_2$S nanorods on conducting Cu substrate in a novel rapid solution based direct synthesis method reflected superior specific of 750 mF cm^{-2} at a current density of 2 mAcm^{-2}. The stable electrodes retained capacitance of 82.3% at a high current density of 40 mAcm^{-2} after 20,000 cycles. An asymmetric device fabricated using activated carbon the AC//SC-Cu$_2$S delivers a higher energy density of 2.69 mW h cm^{-3} and power density ranging from 10.9 to 218.4 mW cm^{-3} in a wide potential window 1.6 V [48]. Binder free electrodes are acknowledged to provide admirable electrochemical activity owing to large surface area, electrode wettability and enhanced conductivity, and hence are being widely investigated [49]. Attempts have also been directed towards commercial-level mass loadings to oversee their applicability. A recent study by Choi et al. using rGO on biomass derived porous activated carbon electrode shows high capacitance of 3.9 F cm^{-2} at density of 1.2 mAcm^{-2} with mass loading of 12mg/cm^2 [50]. In another similar study Manuraj et al. achieved a capacitance of 244 F g^{-1} at 1 A g^{-1} with high mass loading 30mg/cm^2 of MoS$_2$ nanowires onto nickel foam to assemble a symmetric capacitor [51].

6.2.5 Microsupercapacitors

Microsupercapacitors possess necessary ultrahigh power delivery and desired rate capability for high performance outcome. Meanwhile, miniaturization of supercapacitors is conducive to opening new avenues of application in diverse fields [52–54]. In one study an all-solid state supercapacitor using sulfur doped graphene and ionic gel electrolyte could achieve a volume capacitance of ~582 F cm^{-3} at 10 mV/s and 8.1 F cm^{-3} even at the high scan rates of 2000 V/s with a power density of ~1191 W cm^{-3} [55]. Orangi et al. recently studied 3D printed additive free 2D Ti3C$_2$T$_x$Mxenes with interdigital architectures and obtained an areal capacitance of 1035 mF cm^{-2} at 2 mV/s while achieving a high energy density of 51.7 μW h cm^{-2}) [56]. In a very recent study, FAT-Mxene micro-supercapacitors were assembled using Mxene nanosheets from m-Mxene via simple water freexing method. A remarkable volumetric capacitance of 23.6 mF cm^{-2} and 591 F cm^{-3} was achieved. The energy density was in the range of 10.3 to 29.6 mWh cm^{-3} with power density variation from 18.6 to 3.1 W cm^{-3} elucidating the superior electrochemical performance which stood even higher than lithium film batteries and commercially available supercapacitors [57].

6.2.6 Hybrid Supercapacitors

A hybrid supercapacitor or supercapattery serves as an effective way to overcome the lower energy density of the supercapacitor by employing battery-type material as one of the electrodes. The incorporation of different mechanisms enable high power density (~0.1–30kW kg^{-1}), sizeable energy density (~5–200 Wh kg^{-1}) with excellent stability. As an emerging field it is assumed to be a major stakeholder in the commercial market in the coming years and has attracted immense R&D [17, 58]. Wang and group used amorphous carbon (DC) as negative electrode and mesoporous graphene (MG) as positive electrode and successfully assembled a hybrid battery-supercapacitor (sodium ion). The hybrid ion capacitor could reach energy density of 168 Wh kg^{-1} at 501 W kg^{-1} along with power densities of 2432 W kg^{-1} [59]. With the advent of different polyvalent ions based batteries as alternatives to Li-ion batteries in recent years, focus has also been targeted towards utilisation of same storage theories in field of hybrid supercapacitor [58]. In an interesting study, a new type of Zn-ion hybrid supercapacitor was assembled which could provide an excellent electrochemical performance displaying energy density as high as 84 Wh kg^{-1} at 14.9kW kg^{-1} power density assisted with ultra-long stability of 91% even after 10,000 cycles [60].

6.2.7 Flexible, Wearable and Smart Supercapacitors

Flexible, wearable, smart supercapacitors have gained tremendous popularity among researchers given their deployment in next-generation smart electronics and applications. The shape constraint barrier removal enables their use in limited space with ample electrochemical activity for sustainable technologies [61, 62]. A flexible all-solid-state supercapacitor assembled by Zhao *et al.* using nitrogen, phosphorus, and oxygen doped graphene achieved considerable mass and volume energy densities 25.3 Wh kg^{-1} and 25.2 Kg^{-1} [63]. Textile supercapacitor for sportswear, work fabrics kind of advanced technology has enormous market prospects. In this regard, MnO_2/AC //ACT wearable supercapacitor on cotton substrate was synthesized by Bao and group and displayed extremely high maximum energy density and power density of 66.7 Wh kg^{-1} and 4.97 kW kg^{-1} [64]. In a very exciting novel study the possibility of completely transparent with high flexibility was presented by Xu *et al.* The free standing MnO_2/Ni network exhibited areal capacitance 80.7 mF/cm^2 at 5mV/s which could undergo any kind of deformation with negligible performance attenuation [65].

6.3 Advancements in Battery Technology

Battery is the device which converts chemical energy to electrical energy by means of electrochemical oxidation reaction. A battery is the combination of one or more cell in a series which undergo the chemical reactions to create the flow of electrons within the circuit. A battery comprises mainly three components, i.e., negative electrode (Anode), positive electrode (cathode) and electrolyte. A negative electrode produce electron to the external circuit to which the battery is connected which creates a potential difference between the electrodes. As a result, electron try to redistribute themselves which is controlled by the electrolyte. The circuit connected with batteries provides a path for the electron to move from anode to cathode. On the basis of reusability and rechargeability, batteries are classified into two types, i.e., (a) primary batteries: it is an electrochemical cell which cannot be recharged or reused once depleted and also electrochemical reaction cannot be reversed. The different forms of primary batteries range from AA to coin cell where charging is impossible. Primary cells are basically used in pacemakers, wrist watches, remote, animal trackers, etc. Among all primary batteries alkaline batteries draw great attention because of their cost-effectiveness, environmental friendliness, with high specific energy, etc., which can be easily transferrable and can be stored for a long period. (b) Secondary batteries: it is a type in which the chemical reaction can be reversed by applying requisite amount of voltages in the reverse direction. As it can be recharged again and again for a long period after being used up it is also known as rechargeable batteries. Secondary batteries are used in portable electronic devices like power bank, mobile phones, laptops, electric vehicles and in various electronic gadgets. But still it is a challenge for the scientific community to develop cost-effective materials that can provide more energy and be suitable for storing charge for the long time. The detailed classification of batteries that have been discussed is shown in Figure 6.5.

Here we mainly focus on the secondary batteries that are commercially available on the market and economically viable products like Nickel cadmium cell (Ni-Cd), Nickel-Metal hydride (Ni-M), lead storage, sodium-sulphur, flow batteries and Lithium ion batteries (LIBs). Here, we are dealing with the most recent developed design, operating mechanism and construction of cell using electrode materials and their advantages and disadvantages. In the current scenario the technological innovation and newly developed batteries with high efficiency, energy density, and long life cycle with effective recyclability have improved significantly [66].

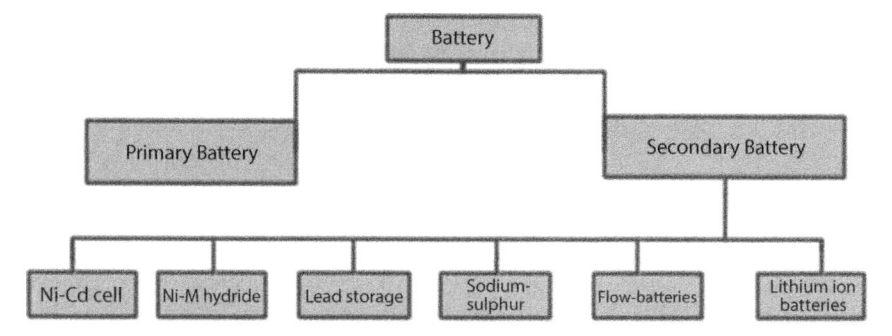

Figure 6.5 Classification of batteries.

6.3.1 Challenges

Till date ample research has been conducted regarding the development of newly designed of batteries with numerous flexibility. However, a number of challenges have to be overcome for their practical application like easy degradation of electrodes under external load, poor performances, complicated cell fabrication, low energy density and power density. Lithium ion battery plays an important role in the energy storage devices but due to rigid electrode materials it is restricted to some extent in direct application [67, 68]. In this regard, R&D and scientific community mitigate the above said issues/challenge by exploiting the proper electrode materials and their cell structure, maintaining the high energy density under industrial electrode manufacturing condition. In view of this, recently developed and in-depth understanding of the current challenges of batteries has been explored for the benefit of scientific readers.

6.3.2 Nickel-Cadmium Batteries

Nickel Cadmium batteries are rechargeable batteries and are made up of Nickel oxide hydroxide and cadmium as electrode material. It was first developed by Swedish scientist Waldemar Junger in 1899. One of the most important features of this cell is that it maintains the voltages and holds charges for a long time when it is not in use. This type of battery gained importance as a reliable, high energy density, large temperature range, lifelong electrochemical system for use in various applications. In Ni-Cd battery nickel oxyhydroxide as cathode and metallic cadmium as anode and alkaline electrolyte such as potassium hydroxide. During the course of discharge it transforms chemical energy into electrical and during recharge it converts electrical energy to chemical energy. However, it has various disadvantages concerning improper charging and toxic metal cadmium discharged into the environment when the life of the battery is over [69, 70]. The electrochemical cell representation and chemical reaction taking place during recharging and discharging in Ni-Cd battery is as follows:

$$2NiO(OH) + Cd + 2H_2O \underset{Discharging}{\overset{Charging}{\rightleftharpoons}} 2Ni(OH)_2 + Cd(OH)_2$$

6.3.3 Nickel-Metal Hydride Batteries

It is another type of rechargeable battery where the chemical reaction taking place is similar to that of the Ni-Cd battery. In this battery the negative electrode uses a hydrogen absorbing alloy instead of cadmium. The energy density and capacity of this cell is almost two to three times better than that of the Ni-Cd battery. Because of their high capacity and energy density it is used in the high drain devices. The Ni-M hydride battery possesses various properties like specific energy 60-120 h/kg, energy density 140-300 Wh/L, specific power 250-1000 W/kg with efficient charging-discharging property with a high degree of cyclic stability. Cheng Tan and co-workers [71] developed negative electrode materials for Ni-MH which delivered excellent electrochemical property. They have reported the selective entry of Sm in Sm-dopedPr/Nd/Mg-free low cobalt La3.0xCexSm0.98-4xZr0.02Ni3.91Co0.14Mn0.25Al0.30 (x ¼ 0.08; 0.12; 0.16; 0.20; 0.245) as electrode materials. The incorporation of Sm results in the enhancement of electrochemical properties like maximum discharge capacity of 331.2 mAhg^{-1} with excellent retention capacity of 78.30%. Taichi Iwai *et al.* [72] investigated the effect of local cell reaction between active materials and current collector. They evaluated the degradation of battery and the cell performance by employing the various metal as current collector as on local cell. The higher potential materials (Pt and Au) as current collector will work to prevent the phase transformation and also the degradation of the battery. However, their practical use in large scale is not familiar due to the short degradation, and also the cost of the materials is also not economically viable due to the presence of Pt and Au.

6.3.4 Lead Storage Battery

Lead storage cell is mostly used in heavy duty applications due to the reliable power sources and low-cost. The size and weight of this battery is very high; therefore it cannot be used in portable electronic devices. However,

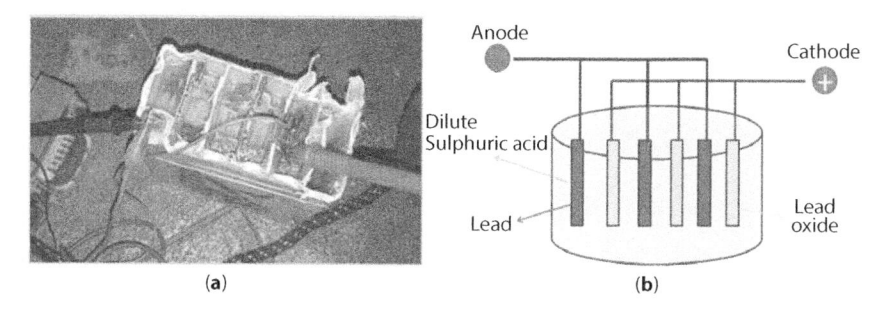

Optimistic Technological Approaches for Sustainable Energy 215

Figure 6.6 Internal view and construction of lead-storage cell.

the lead storage battery is one of the oldest rechargeable batteries and now also plays an important role in day-to-day life. It has very low energy to volume ratio as well as energy to weight ratio and has the capability to produce huge surge current when needed for a short period of time. Due to these properties it is used in the solar-panel, ignition of vehicles and their lights, backup power and load leveling in power generation/distribution. The internal view and construction of the lead storage battery is shown in Figure 6.6 (a & b). The chemical reaction taking place in lead storage is as given below.

$$\text{Reaction at anode: Pb} \longrightarrow Pb^{2+} 2^{e-1}$$

$$\text{Reaction at cathode: } PbO_2 + 4H^+ + 2e^- \longrightarrow Pb^{2+} + 2H_2O$$

$$\text{Cell reaction: } Pb + PbO_2^+ + 4H^+ + 2SO_4^{2-} \longrightarrow 2PbSO_4 + 2H_2O$$

6.3.5 Sodium Sulphur Battery

It is another type of secondary batteries which was developed by Ford Motor Company in 1960 and in the market it is sold by the Japanese company NGK. This battery uses solid electrolyte (sodium beta alumina) and the operating temperature of 300 °C to 350 °C. It is also known as molten-salt battery where liquid sodium and sulphur are used in positive electrode and negative electrode and is separated by solid electrolyte, the chemical reaction between two taking place by cell reaction (Figure 6.7). The current is generated sodium metal release the electrons and forms sodium ions moving to the positive electrode through electrolyte.

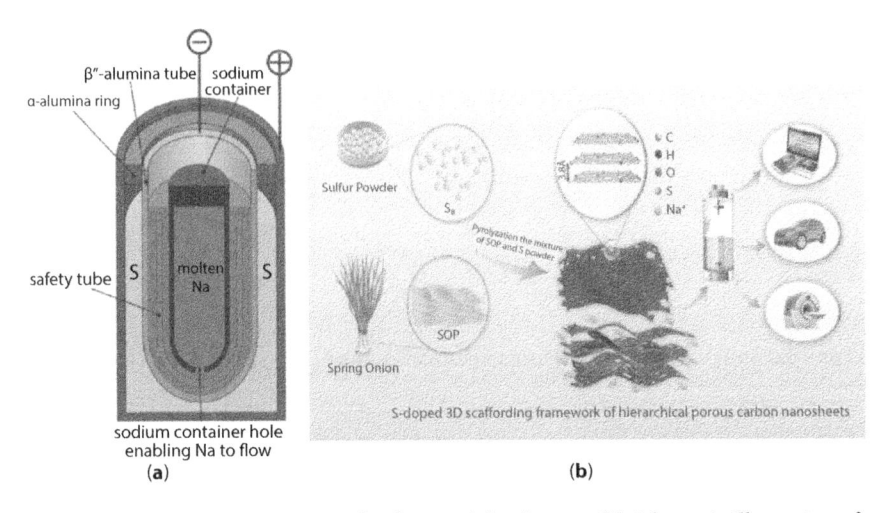

Figure 6.7 (a) inner construction of sodium-sulphur battery (b) Schematic illustration of the synthesis of the 3D scaffolding framework of S-doped and their application in battery. Reprint with permission [73, 74].

Generally, the specific density is about 150 Khkg^{-1} or sometimes higher than this. The cell is mainly cylindrical in shape made up of steel whose inner surface is coated with chromium and molybdenum to prevent the entire cell from corrosion. During the discharge sodium metal loses its electron and migrates to the sulphur electrode material. The free electron drives an electric current through molten sodium to the contact through the electrical load and back to sulphur container. The chemical reaction in the discharge process is given below.

$$2\,Na + 4\,S \longrightarrow Na_2S_4$$

Nowadays researchers have replaced the conventional solid electrolyte from glass fiber separator soaked ether-based electrolyte that allows it to be used in room temperature. Gongyuan Zhao *et al.* [73] developed powerful electrode materials by using the mixture of biomass and sulphur powder as potential electrode material for scalable commercial production of sodium sulphur battery. The electrochemical performances were carried out by using three electrode systems by using 1 mol L^{-1} NaClO$_4$ as electrolytic solution. The electrolyte is made up by dissolving NaClO$_4$ in ethylene carbonate and diethyl carbonate at a volume ratio 1:1 with 5 wt% fluoroethylene carbonate. The electrode material possess ultrahigh electrochemical properties of reversible capacity of 605 mAhg^{-1} at current density of 0.05 Ag⁻¹ and long cyclic stability. The cyclic retention value is about 94% after the completion of 2000 cycles at current density 5 Ag⁻¹.

6.3.6 Flow Batteries

It is a type of rechargeable fuel cell in which an electrolyte containing one or more dissolved electroactive element flows through electrochemical cell that reversibly converts chemical energy to electricity. The electroactive element means an element in solution that takes part in an electrode reaction or can be easily absorbed by the electrode materials. Flow batteries are essential for integrating variable renewable energy source into electricity grid. In this cell electrolytes are kept externally in tanks and are usually pumped through cell or reactor. Flow batteries comprise three components, i.e, cell stack (CS), electrolyte storage (ES) and auxiliary parts or balance-of-plant where cell stack determines the power rating for the system and is assembled from several cells stacked together. The stack is supported by several components like current collectors, gaskets, and stack shells [75]. The energy capacity of the flow battery can be determined by liquid electrolytes that are stored in tanks [76]. Because of their unique properties like decoupled power and capacity it can be treated as one of the most promising energy devices for grid-scale application [77]. Ample research has been done by the scientific community to develop suitable electrode materials for flow batteries which can be treated as future material for their practical application. Andrew W. Lantz and co-workers [78] synthesized sulphonated derivatives of 4,4'- biphenol and 1,8-dihydroxyanthraquinone and evaluated their electrochemical behaviour towards the application in flow battery. The electrochemical evaluation was carried out by using three electrode system possessing long-term charging/discharging measurement to determine the energy efficiency, charge capacity and stability, etc. The synthesized electrolyte with standard cell potential of 0.905 V with a current density of 0.8 Acm⁻² having energy density of 45 WhL⁻¹ using 1 M electrolytic solution. After the completion of this reaction the coulombic efficiency of the flow battery stabilizes 95%. Thus it can be considered as one of the most promising electrolytes for the development of flow battery in future endeavors. Craig G. Armstrong *et al.* [79] reported the ferrocene/ferrocenium (Fc/FcBF$_4$) as a low-cost model of chemistry for non-aqueous flow battery. The electrochemical properties were fully characterized by cyclic voltammetry, charging-discharging, and cyclic cyclic stability. They explored it as an important material based on the single redox couple cycling which possesses high capacity retention of ferrocene at 10 mM concentration with the capacity retention of 80% after the completion of

200 cycles. This small loss of retention capacity was identified and reported by them which may occur due to the chemical instability of Fc/FcBF$_4$ and their oxidation state. Kathryn E. Toghill and co-workers [80] synthesized cobalt complexes with azole pyridine ligands for non-aqueous redox-flow batteries and their tunable electrochemical properties by structure modification. The Co (II/I) and Co (III/II) couples are stable and redox potential are tunable allowing the potential difference increase from 1.07 to 1.91 V via pyridine substitution with weaker σ-donating/π-accepting azole groups. It possesses high coulombic efficiencies of 89.7% - 99.8% with very good voltaic efficiencies of 70.3 - 81.0%. The modification of ligands leads via pyridyl substitution with 3,5-dimethylpyrazole to improve the solubility from 0.18M to 0.5M via substitution with azole group.

6.3.7 Lithium Ion Batteries (LIBs)

Among all the above-discussed rechargeable batteries a Lithium ion battery is one of the most advanced rechargeable batteries, and it has drawn great attention in the past few decades. In the current scenario lithium ion batteries are an integral part of our life, powering portable devices like cellphones and laptops that have revolutionized the whole world [81, 82]. LIBs are considered as the powerhouse of personal electronic and optoelectronic devices that revolutionized the electronic world at the same time the commercialization of LIBs were also started. It is extensively used in a wide range of applications owing to their high coulombic efficiencies, high energy densities, low self-discharge features, and also a huge range of chemical potential which is accessible as suitable electrode materials. The increasing demands for high-performance rechargeable batteries is seen everywhere in this modern society, needed in order to develop the new and advanced material which can meet the demands of the market. In the last few decades the scientific community and researchers have seen the development of new electrode materials for the next-generation lithium-ion batteries which can provide high charge capacities/power densities, high charging rate, long life and low-cost for electric vehicles (EV), hybrid electric vehicles, aerospace applications, and autonomous electric devices. However, the demand for and the life of the battery restricts the development of electric vehicles. The whole life cycle of the battery depends on production, design and EV application as shown in Figure 6.8(a). The battery aging mechanism and the impact of battery degradation should be considered to be optimized during the designing of the battery. The battery aging analysis is generally carried out on several influencing factors like

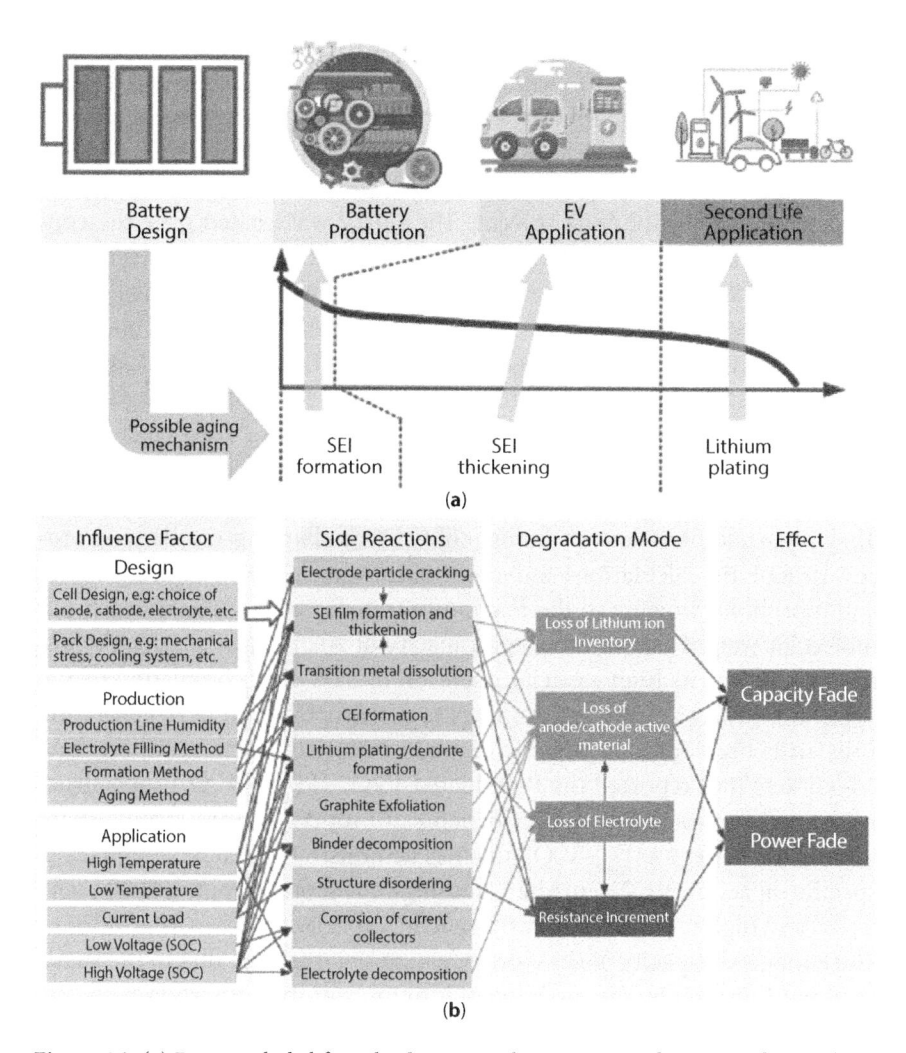

Figure 6.8 (a) Battery whole life cycle: design, production, EV application and second life application, and (b) Cause and effect of degradation mechanisms and associated degradation modes. Reprint with permission [83].

internal side reactions, degradation modes and external affects as shown in Figure 6.8 (b) [83].

Therefore it is still a challenge for the scientific community to develop new materials, methods and design which will meet demand and serve as potential materials. In lieu of this, various research has been reported by the scientific community to overcome these issues and make the battery commercially viable with long life cycles. Qiang Li *et al.* [84] synthesized monodispersed hollow Fe_2O_4 nanospheres by using hydrothermal method. The electrochemical studies were carried out in CR2032 coin type battery geometry where lithium metal was taken as counter electrode. For electrochemical evaluation the electrode material was prepared by mixing hollow Fe_2O_4 with conductive carbon black and PVDF in the weight ratio 7:2:1 in an N-methyl-2-pyrrolidone solvent. The slurry was coated over the copper foil and allows it to dry overnight to remove the residual solvent. The electrochemical discharge capacity was recorded to be 1718 mAhg^{-1} in first cycle and is stabilized to 1370 mAhg^{-1} and 1364mAhg^{-1} in second and third cycle, respectively, which is quite higher than that of the theoretical value. Hence, in future endeavors this material will provide fundamental guidance for the development of advanced energy storage systems. Yong Lu et al. [85] reported the successfully synthesized organic carbonyl compounds cyclohexanehexone (C_6O_6) and investigated its application in the field of LIBs. They reported the redox mechanism of C_6O_6 by charging-discharging process using DFT calculations and ex-situ techniques. The structural features and morphology owing to the maximum transferred electron number per molecular weight exhibit ultrahigh capacity of 902 mAhg^{-1} at current density 20 mAg^{-1}. The high capacity retention of 82% after completion of 100 cycles at 50 mAg^{-1} gives opportunities to construct/design next-generation LIBs with high-capacity organic cathode materials. Wen WuZhong and co-workers [86] reported the Prelithiated V2O5 ($Li_{0.0625}V_2O_5$) superstructures was synthesized and electrochemical properties were evaluated for lithium ion battery. $Li_{0.0625}V_2O_5$ moieties demonstrated that it has enhanced specific capacity upto 215 mAhg^{-1}, ultrahigh cyclic stability after 1000 cycles at 1 C and high rate capability with specific capacity of 140 mAhg^{-1} at 20 C. The enhanced capacity of the electrode material might be due to the structural modification by the insertion of lithium over the orthorhombic V_2O_5, crystal structure stability and improved conductivity.

The market for secondary battery has drawn great attention after the rapid globalization/invention of portable electrical devices such as portable electronic calculator, implantable electronic or simple flashlights [87]. There have been many batteries utilized for the smooth running of these devices but their size, weight and specific capacity and life span restricts them. In contrast, LIBs play an important role in the replacement of all old batteries. It was expected that nearly 100 GW hours of LIBs are required to fulfill the needs for consumer use and electric–powered vehicles with the latter taking about 50% of LIBs sale by 2018 [81]. The lithium ion batteries can be used in large scale in grid application and will require next-generation batteries to be produced at low cost. It was expected that in 2020 the

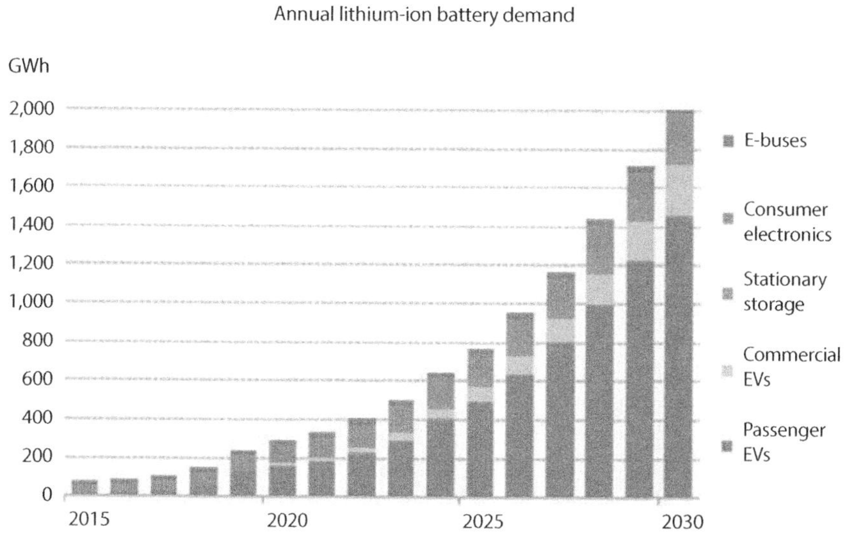

Annual lithium-ion battery demand

Figure 6.9 The expected annual lithium ion battery demand from 2015 to 2030.

energy required to fulfill the need would be around 300 GW hours. As per the report of Energy Collective Group the expected annual demand of lithium batteries per year is presented in Figure 6.9. According to their report, the energy demands/consumption in terms of LIBs will be around 2000 GWh in 2030 by E-buses, electronic devices used by global consumers, stationary storage, commercial electric vehicles, and private electric vehicles, etc. The demand of LIBs will increase by 6-7 times in the coming decades compared to the situation today.

6.4 Conclusion and Outlook

Batteries and supercapacitors are versatile and green energy storing sources with immense potential of realising the vision of a sustainable society. Both batteries and supercapacitors have their advantages and limitations that determine their end applications. There have been significant advancements in both fields, with an an eye on the huge opportunities in the years to come. However, there still are a lot of constraints to overcome in terms of performance, manufacturing and commercialisation. The landscape of utilising both the energy delivery of batteries and power delivery of supercapacitors offers enticing forecasts to be utilised. In the current scenario, with increasing popularity of smart electronic, electric vehicles a transition towards a green society can be seen; development of novel materials and improved electrochemical performance can greatly help to facilitate the progress. To conclude, Supercapacitors and batteries offer tremendous prospects that will surely drive tremendous research exploration and efforts toward performance enrichment for a broad spectrum of applications.

References

1. Simon, P. & Gogotsi, Y. Materials for electrochemical capacitors. *Nat. Mater.* 7, 845–854 (2008).
2. Liu, C. F., Liu, Y. C., Yi, T. Y. & Hu, C. C. Carbon materials for high-voltage supercapacitors. *Carbon* 145, 529–548 (2019).
3. Wu, H. *et al.* Graphene based architectures for electrochemical capacitors. *Energy Storage Mater.* 5, 8–32 (2016).
4. Horn, M., MacLeod, J., Liu, M., Webb, J. & Motta, N. Supercapacitors: a new source of power for electric cars? *Econ. Anal. Policy* 61, 93–103 (2019).
5. Kötz, R., & Carlen, M. J. E. A. (2000). Principles and applications of electrochemical capacitors. *Electrochimicaacta*, 45(15–16), 2483–2498.
6. Jia, R., Shen, G., Qu, F., & Chen, D. (2020). Flexible on-chip micro-supercapacitors: Efficient power units for wearable electronics. *Energy Storage Materials*, 27, 169–186.
7. Miao, L., Song, Z., Zhu, D., Li, L., Gan, L., & Liu, M. (2020). Recent advances in carbon-based supercapacitors. *Materials Advances*, 1(5), 945–966.
8. Panda, P. K., Grigoriev, A., Mishra, Y. K., & Ahuja, R. (2020). Progress in supercapacitors: roles of two dimensional nanotubular materials. *Nanoscale Advances*, 2(1), 70–108.
9. Zhang, F., Zhang, T., Yang, X., Zhang, L., Leng, K., Huang, Y., & Chen, Y. (2013). A high-performance supercapacitor-battery hybrid energy storage device based on graphene-enhanced electrode materials with ultrahigh energy density. *Energy & Environmental Science*, 6(5), 1623–1632.
10. Shi, Y., Peng, L., Ding, Y., Zhao, Y., & Yu, G. (2015). Nanostructured conductive polymers for advanced energy storage. *Chemical Society Reviews*, 44(19), 6684–6696.
11. Li, B., Dai, F., Xiao, Q., Yang, L., Shen, J., Zhang, C., &Cai, M. (2016). Nitrogen-doped activated carbon for a high energy hybrid supercapacitor. *Energy & Environmental Science*, 9(1), 102–106.
12. Zhu, G., He, Z., Chen, J., Zhao, J., Feng, X., Ma, Y., & Huang, W. (2014). Highly conductive three-dimensional MnO_2–carbon nanotube–graphene–Ni hybrid foam as a binder-free supercapacitor electrode. *Nanoscale*, 6(2), 1079–1085.

13. Conway, B. E. (2013). *Electrochemical supercapacitors: scientific fundamentals and technological applications*. Springer Science & Business Media.
14. Brousse, T., Bélanger, D., & Long, J. W. (2015). To be or not to be pseudocapacitive. *J. Electrochem. Soc.*, 162(5), A5185–A5189.
15. Das, S., & Ghosh, A. (2020). Symmetric electric double-layer capacitor containing imidazolium ionic liquid-based solid polymer electrolyte: Effect of TiO_2 and ZnO nanoparticles on electrochemical behavior. *Journal of Applied Polymer Science*, 137(22), 48757.
16. Dong, M., Wang, Z., Yan, G., Wang, J., Guo, H., & Li, X. (2020). Confine growth of $NiCo_2S_4$ nanoneedles in graphene framework toward high-performance asymmetric capacitor. *Journal of Alloys and Compounds*, 822, 153645.
17. Balasubramaniam, S., Mohanty, A., Balasingam, S. K., Kim, S. J., & Ramadoss, A. (2020). Comprehensive Insight into the Mechanism, Material Selection and Performance Evaluation of Supercapatteries. *Nano-Micro Letters*, 12, 1–46.
18. Shi, X., Zhang, S., Chen, X., Tang, T., & Mijowska, E. (2020). Three dimensional graphene/carbonized metal-organic frameworks based high-performance supercapacitor. *Carbon*, 157, 55–63.
19. Largeot, C., Portet, C., Chmiola, J., Taberna, P. L., Gogotsi, Y., & Simon, P. (2008). Relation between the ion size and pore size for an electric double-layer capacitor. *Journal of the American Chemical Society*, 130(9), 2730–2731.
20. Lokhande, A. C., Teotia, S., Shelke, A. R., Hussain, T., Qattan, I. A., Lokhande, V. C., & Lokhande, C. D. (2020). Chalcopyrite based carbon composite electrodes for high performance symmetric supercapacitor. *Chemical Engineering Journal*, 125711.
21. Zhou, Y., Guo, W., & Li, T. (2019). A review on transition metal nitrides as electrode materials for supercapacitors. *Ceramics International*, 45(17), 21062–21076.
22. Lyu, L., Seong, K. D., Ko, D., Choi, J., Lee, C., Hwang, T., & Piao, Y. (2019). Recent development of biomass-derived carbons and composites as electrode materials for supercapacitors. *Materials Chemistry Frontiers*, 3(12), 2543–2570.
23. Ho, K. C., & Lin, L. Y. (2019). A review of electrode materials based on core–shell nanostructures for electrochemical supercapacitors. *Journal of Materials Chemistry A*, 7(8), 3516–3530.
24. Najib, S., & Erdem, E. (2019). Current progress achieved in novel materials for supercapacitor electrodes: mini review. *Nanoscale Advances*, 1(8), 2817–2827.
25. Yang, Y. (2020). A mini-review: emerging all-solid-state energy storage electrode materials for flexible devices. *Nanoscale*, 12(6), 3560–3573.
26. Pal, B., Yang, S., Ramesh, S., Thangadurai, V., & Jose, R. (2019). Electrolyte selection for supercapacitive devices: a critical review. *Nanoscale Advances*, 1(10), 3807–3835.

27. Dou, Q., Lei, S., Wang, D. W., Zhang, Q., Xiao, D., Guo, H., & Yan, X. (2018). Safe and high-rate supercapacitors based on an "acetonitrile/water in salt" hybrid electrolyte. *Energy & Environmental Science*, 11(11), 3212–3219.
28. Ngai, K. S., Ramesh, S., Ramesh, K., & Juan, J. C. (2016). A review of polymer electrolytes: fundamental, approaches and applications. *Ionics*, 22(8), 1259–1279.
29. Supercapacitor Market Report: Trends, Forecast and Competitive Analysis. https://www.researchandmarkets.com/reports/4912100/supercapacitor-market-report-trends-forecast. 2020.
30. Supercapacitor Market by Product Type, Module Type, Material and Application: Global Opportunity Analysis and Industry Forecast, 2020-2027. https://www.researchandmarkets.com/reports/5134092/supercapacitor-market-by-product-type-module. 2020.
31. Supercapacitors Market - Growth, Trends, and Forecasts (2020 - 2025). https://www.researchandmarkets.com/reports/4745444/supercapacitors-market-growth-trends-and. 2020.
32. Maxwell Technologies, http://www.maxwell.com/products/ultracapacitors. 2019.
33. Li, H., Hou, Y., Wang, F., Lohe, M. R., Zhuang, X., Niu, L., & Feng, X. (2017). Flexible all-solid-state supercapacitors with high volumetric capacitances boosted by solution processable MXene and electrochemically exfoliated graphene. *Advanced Energy Materials*, 7(4), 1601847.
34. Guo, Y., Li, W., Yu, H., Perepichka, D. F., & Meng, H. (2017). Flexible asymmetric supercapacitors via spray coating of a new electrochromic donor–acceptor polymer. *Advanced Energy Materials*, 7(2), 1601623.
35. Scalia, A., Bella, F., Lamberti, A., Bianco, S., Gerbaldi, C., Tresso, E., & Pirri, C. F. (2017). A flexible and portable powerpack by solid-state supercapacitor and dye-sensitized solar cell integration. *Journal of Power Sources*, 359, 311–321.
36. VINATech. Supercapacitor.Solution. Jan. http://www.vina.co.kr/eng/product/supercap.html. 2019.
37. Yunasko. Ultracapacitor Product Line. http://yunasko.com/en/products.2019.
38. Rinne, J., Keskinen, J., Berger, P. R., Lupo, D., & Valkama, M. (2017). Viability bounds of M2M communication using energy-harvesting and passive wake-up radio. *IEEE Access*, 5, 27868-27878.
39. Adifproject. http://www.adif.es/es_ES/comunicacion_y_prensa/%1Cchas_de_actualidad/%1Ccha_actualidad_00072.shtml
40. Hashemi, M., Rahmanifar, M. S., El-Kady, M. F., Noori, A., Mousavi, M. F., & Kaner, R. B. (2018). The use of an electrocatalytic redox electrolyte for pushing the energy density boundary of a flexible polyaniline electrode to a new limit. *Nano Energy*, 44, 489–498.
41. Rawool, C. R., Punde, N. S., Rajpurohit, A. S., Karna, S. P., & Srivastava, A. K. (2018). High energy density supercapacitive material based on a ternary

hybrid nanocomposite of cobalt hexacyanoferrate/carbon nanofibers/polypyrrole. *ElectrochimicaActa*, 268, 411–423.
42. Dura, H., Perry, J., Lecacou, T., Markoulidis, F., Lei, C., Khalil, S., ...& Weil, M. (2013, June). Cost analysis of supercapacitor cell production. In *2013 international conference on clean electrical power (ICCEP)* (pp. 516–523). IEEE.
43. NorthernGraphite, Graphite Pricing, (2018). http://northerngraphite.com/graphite-. 2018.
44. Ajina, A. (2015). Statistical optimization of supercapacitor pilot plant manufacturing and process scale-up (Doctoral dissertation, University of Nottingham).
45. AbiJaoude, M., Alhseinat, E., Polychronopoulou, K., Bharath, G., Darawsheh, I. F. F., Anwer, S., ...& Banat, F. (2020). Morphology-dependent electrochemical performance of MnO2 nanostructures on graphene towards efficient capacitive deionization. *ElectrochimicaActa*, 330, 135202.
46. Patil, A. M., Lokhande, A. C., Shinde, P. A., & Lokhande, C. D. (2018). Flexible asymmetric solid-state supercapacitors by highly efficient 3D nanostructured α-MnO2 and h-CuS electrodes. *ACS applied materials & interfaces*, 10(19), 16636–16649.
47. Zhao, F., Xie, D., Huang, W., Song, X., Sial, M. A. Z. G., Wu, H., ...& Zeng, X. (2020). Defect-rich honeycomb-like nickel cobalt sulfides on graphene through rapid microwave-induced synthesis for ultrahigh rate supercapacitors. *Journal of Colloid and Interface Science*, 580, 160–170.
48. Hong, J., Kim, B. S., Yang, S., Jang, A. R., Lee, Y. W., Pak, S., ...& Hong, J. P. (2019). Chalcogenide solution-mediated activation protocol for scalable and ultrafast synthesis of single-crystalline 1-D copper sulfide for supercapacitors. *Journal of Materials Chemistry A*, 7(6), 2529–2535.
49. Gopalakrishnan, A., Sahatiya, P., & Badhulika, S. (2018). Template-assisted electrospinning of bubbled carbon nanofibers as binder-free electrodes for high-performance supercapacitors. *ChemElectroChem*, 5(3), 531–539.
50. Choi, J. H., Kim, Y., & Kim, B. S. (2020). Multifunctional role of reduced graphene oxide binder for high performance supercapacitor with commercial-level mass loading. *Journal of Power Sources*, 454, 227917.
51. Manuraj, M., Nair, K. K., Unni, K. N., & Rakhi, R. B. (2020). High performance supercapacitors based on MoS_2 nanostructures with near commercial mass loading. *Journal of Alloys and Compounds*, 819, 152963.
52. Wan, S., Peng, J., Jiang, L., & Cheng, Q. (2016). Bioinspired graphene-based nanocomposites and their application in flexible energy devices. *Advanced Materials*, 28(36), 7862–7898.
53. Huang, P., Lethien, C., Pinaud, S., Brousse, K., Laloo, R., Turq, V., ...& Chaudret, B. (2016). On-chip and freestanding elastic carbon films for micro-supercapacitors. *Science*, 351(6274), 691–695.

54. Zhang, C. J., McKeon, L., Kremer, M. P., Park, S. H., Ronan, O., Seral-Ascaso, A., ...& Anasori, B. (2019). Additive-free MXene inks and direct printing of micro-supercapacitors. *Nature Communications*, 10(1), 1–9.
55. Wu, Z. S., Tan, Y. Z., Zheng, S., Wang, S., Parvez, K., Qin, J., ...& Mullen, K. (2017). Bottom-up fabrication of sulfur-doped graphene films derived from sulfur-annulated nanographene for ultrahigh volumetric capacitance micro-supercapacitors. *Journal of the American Chemical Society*, 139(12), 4506–4512.
56. Orangi, J., Hamade, F., Davis, V. A., & Beidaghi, M. (2019). 3D Printing of Additive-Free 2D Ti3C2T x (MXene) Ink for Fabrication of Micro-Supercapacitors with Ultra-High Energy Densities. *ACS nano*, 14(1), 640–650.
57. Huang, X., & Wu, P. (2020). A Facile, High-Yield, and Freeze-and-Thaw-Assisted Approach to Fabricate MXene with Plentiful Wrinkles and Its Application in On-Chip Micro-Supercapacitors. *Advanced Functional Materials*, 30(12), 1910048.
58. Zuo, W., Li, R., Zhou, C., Li, Y., Xia, J., & Liu, J. (2017). Battery-supercapacitor hybrid devices: recent progress and future prospects. *Advanced Science*, 4(7), 1600539.
59. Wang, F., Wang, X., Chang, Z., Wu, X., Liu, X., Fu, L. & Huang, W. (2015). A quasi-solid-state sodium-ion capacitor with high energy density. *Advanced Materials*, 27(43), 6962–6968.
60. Dong, L., Ma, X., Li, Y., Zhao, L., Liu, W., Cheng, J., & Kang, F. (2018). Extremely safe, high-rate and ultralong-life zinc-ion hybrid supercapacitors. *Energy Storage Materials*, 13, 96-102.
61. Chee, W. K., Lim, H. N., Zainal, Z., Huang, N. M., Harrison, I., & Andou, Y. (2016). Flexible graphene-based supercapacitors: a review. *Journal of Physical Chemistry C*, 120(8), 4153–4172.
62. Dong, L., Xu, C., Li, Y., Huang, Z. H., Kang, F., Yang, Q. H., & Zhao, X. (2016). Flexible electrodes and supercapacitors for wearable energy storage: a review by category. *Journal of Materials Chemistry A*, 4(13), 4659–4685.
63. Chen, T., Hao, R., Peng, H., & Dai, L. (2015). High-performance, stretchable, wire-shaped supercapacitors. *AngewandteChemie* International Edition, 54(2), 618–622.
64. Bao, L., & Li, X. (2012). Towards textile energy storage from cotton T-shirts. *Advanced Materials*, 24(24), 3246–3252.
65. Liu, Y. H., Xu, J. L., Gao, X., Sun, Y. L., Lv, J. J., Shen, S., & Wang, S. D. (2017). Freestanding transparent metallic network based ultrathin, foldable and designable supercapacitors. *Energy & Environmental Science*, 10(12), 2534–2543.
66. Fan, X., Liu, B., Liu, J., Ding, J., Han, X., Deng, Y., & Zhong, C. (2020). Battery technologies for grid-level large-scale electrical energy storage. *Transactions of Tianjin University*, 1–12.

67. Davies, D. M., Verde, M. G., Mnyshenko, O., Chen, Y. R., Rajeev, R., Meng, Y. S., & Elliott, G. (2019). Combined economic and technological evaluation of battery energy storage for grid applications. *Nature Energy*, 4(1), 42–50.
68. Cha, H., Kim, J., Lee, Y., Cho, J., & Park, M. (2018). Issues and Challenges Facing Flexible Lithium‐Ion Batteries for Practical Application. *Small*, 14(43), 1702989.
69. Boddula, R., Pothu, R., & Asiri, A. M. (Eds.). (2020). *Rechargeable Batteries: History, Progress, and Applications*. John Wiley & Sons.
70. Kurzweil, P., & Scheuerpflug, W. (2020). State-of-Charge Monitoring and Battery Diagnosis of NiCd Cells Using Impedance Spectroscopy. *Batteries*, 6(1), 4.
71. Tan, C., Ouyang, L., Chen, M., Jiang, W., Min, D., Liao, C., & Zhu, M. (2020). Effect of Sm on performance of Pr/Nd/Mg-free and low-cobalt AB4. 6 alloys in nickel-metal hydride battery electrode. *Journal of Alloys and Compounds*, 154530.
72. Iwai, T., Takai, S., Yabutsuka, T., & Yao, T. (2019). Effect of local cell reaction at cathode on the performance of nickel metal-hydride battery. *Journal of Alloys and Compounds*, 772, 256–262.
73. Zhao, G., Yu, D., Zhang, H., Sun, F., Li, J., Zhu, L., & Sun, Y. (2020). Sulphur-doped carbon nanosheets derived from biomass as high-performance anode materials for sodium-ion batteries. *Nano Energy*, 67, 104219.
74. Andriollo, M., Benato, R., Sessa, S. D., Di Pietro, N., Hirai, N., Nakanishi, Y., & Senatore, E. (2016). Energy intensive electrochemical storage in Italy: 34.8 MW sodium–sulphur secondary cells. *Journal of Energy Storage*, 5, 146–155.
75. He, H., Tian, S., Tarroja, B., Ogunseitan, O. A., Samuelsen, S., & Schoenung, J. M. (2020). Flow battery production: Materials selection and environmental impact. *Journal of Cleaner Production*, 121740.
76. Cunha, Á., Martins, J., Rodrigues, N., & Brito, F. P. (2015). Vanadium redox flow batteries: a technology review. *International Journal of Energy Research*, 39(7), 889–918.
77. Armstrong, C. (2020). Novel non-aqueous symmetric redox materials for redox flow battery energy storage (Doctoral dissertation, Lancaster University).
78. Lantz, A. W., Shavalier, S. A., Schroeder, W., & Rasmussen, P. G. (2019). Evaluation of an Aqueous Biphenol-and Anthraquinone-Based Electrolyte Redox Flow Battery. *ACS Applied Energy Materials*, 2(11), 7893–7902.
79. Armstrong, C. G., Hogue, R. W., & Toghill, K. E. (2020). Characterisation of the ferrocene/ferrocenium ion redox couple as a model chemistry for non-aqueous redox flow battery research. *Journal of Electroanalytical Chemistry*, 872, 114241.
80. Armstrong, C. G., & Toghill, K. E. (2017). Cobalt (II) complexes with azole-pyridine type ligands for non-aqueous redox-flow batteries: Tunable electrochemistry via structural modification. *Journal of Power Sources*, 349, 121–129.

81. Deng, D. (2015). Li-ion batteries: basics, progress, and challenges. *Energy Science & Engineering*, 3(5), 385–418.
82. Deng, D., Kim, M. G., Lee, J. Y., & Cho, J. (2009). Green energy storage materials: Nanostructured TiO 2 and Sn-based anodes for lithium-ion batteries. *Energy & Environmental Science*, 2(8), 818–837.
83. Han, X., Lu, L., Zheng, Y., Feng, X., Li, Z., Li, J., & Ouyang, M. (2019). A review on the key issues of the lithium ion battery degradation among the whole life cycle. *ETransportation*, 1, 100005.
84. Li, Q., Li, H., Xia, Q., Hu, Z., Zhu, Y., Yan, S., ...& Fan, S. (2020). Extra storage capacity in transition metal oxide lithium-ion batteries revealed by *in situ* magnetometry. *Nature Materials*, 1–8.
85. Lu, Y., Hou, X., Miao, L., Li, L., Shi, R., Liu, L., & Chen, J. (2019). Cyclohexanehexone with Ultrahigh Capacity as Cathode Materials for Lithium-Ion Batteries. *AngewandteChemie* International Edition, 58(21), 7020–7024.
86. Zhong, W., Huang, J., Liang, S., Liu, J., Li, Y., Cai, G., ...& Liu, J. (2019). New prelithiated V2O5 superstructure for lithium-ion batteries with long cycle life and high power. *ACS Energy Letters*, 5(1), 31–38.
87. Yoshino, A. (2012). The birth of the lithium-ion battery. *AngewandteChemie* International Edition, 51(24), 5798–5800.

7
Electro-Chemical Battery Energy Storage Systems - A Comprehensive Overview

Nikhil P G[1]* and G Sivaramakrishnan[2]

[1]National Institute of Solar Energy, Gurugram, Haryna, India
[2]Chartered Engineer, Institution of Engineers (India) Kolkata, West Bengal, India

---

Abstract
The technology of electro-chemical energy conversion has evolved with time. The concerns are majorly evolving around the implementation aspects of these electro-chemical energy storage systems in the new age application domains. This chapter focuses on the submission of various technology and commercial dimensions of the electro-chemical batteries in the ongoing era. These include energy landscape, storage applications, design basis and performance parameters of an electro-chemical storage, a typical use case from an industrial case study, and overview of recycling aspects.

Keywords: Energy storage landscape, electro-chemical storage

7.1 Introduction

Generation, transmission and distribution, and consumption are the three main pillars of the energy eco-system. Energy storage is emerging as the fourth pillar. As the energy generation resources transition from centralised configurations to a distributed structure, the demanding necessity of storing electrical energy for different applications is becoming prominent. The increasingly felt necessities for optimisation of quantum of energy generated and its efficient utilisation give rise to the need for storage element.

Electricity is the only transportable form of energy and hence has the first right of control in our transition towards meeting Nationally Determined

---

Corresponding author: nikhilpg_mnre@hotmail.com

Umakanta Sahoo (ed.) Energy Storage, (229–252) © 2021 Scrivener Publishing LLC

Contributions. Battery Energy Storage System (BESS) has been in existence for over one and a half centuries. This supports the need for an extensive overview of the various aspects of the emerging electro-chemical storage systems. This chapter is further divided into the following sub-segments:

- a) Drivers, barriers, and benefits in the implementation of Electro-chemical storage devices, and the gist of the energy storage landscape and energy storage applications are discussed in this sub-section.
- b) Batteries can be a crucial component for diesel abatement in the industrial units. Medium and small industries have found a cost-effective solution in energy storage systems. The stored electricity generated out of the intermittent and variable renewable sources are effectively utilized during the peak hour. This sub-segment highlights the various performance parameters and design basis for a battery energy storage system.
- c) Utilities also benefit from the demand side management usage through storage. The peak load shaving of energy reduces the burden of energy imports to meet them. A typical use case from the food industry will be elucidated in this sub-segment.
- d) The maintenance regime in a battery storage system plays a key role in the maximum life cycle for the system. The best practices in the industry along with monitoring and control of key performance indicators of storage systems are captured in the sub-segment.
- e) Electro-chemical Batteries have no resale value and therefore have no secondary market. They have, however, end-of-life value. The chapter concludes with better second life usage of battery systems, disposal of hazardous components after the end of life.
- f) Our National Energy Storage Mission foresees a make in India opportunity for globally competitive manufacturing. The make in India program initiated multiple steps for indigenising this all-important equipment and thus relying less on finished product imports. Policy makers and regulators have created enabling pathways to boost share of stored energy in the network. The major shift in policy includes recognising the role of storage vertical as a key component in the energy transformation. Highlights

of the three-stage solution approach to address the battery manufacturing sector in India are presented in this sub-section.

The submissions in the chapter will be a useful reference for technology practitioners engaged in the battery energy storage industry. The data and knowledge presented shall also add suitable value to the academic fraternity as well.

7.2 Electro-Chemical Storage Devices

This sub-segment of the chapter takes a deep dive into the electro-chemical storage devices, barriers and drivers in the implementation and benefits of such systems. This segment also analyses the energy storage landscape in view of the smart grid implementation of the transmission and distribution sector.

7.2.1 Definition and Types

Energy can be stored in various forms. But the major types of energy storage can be categorised as following:

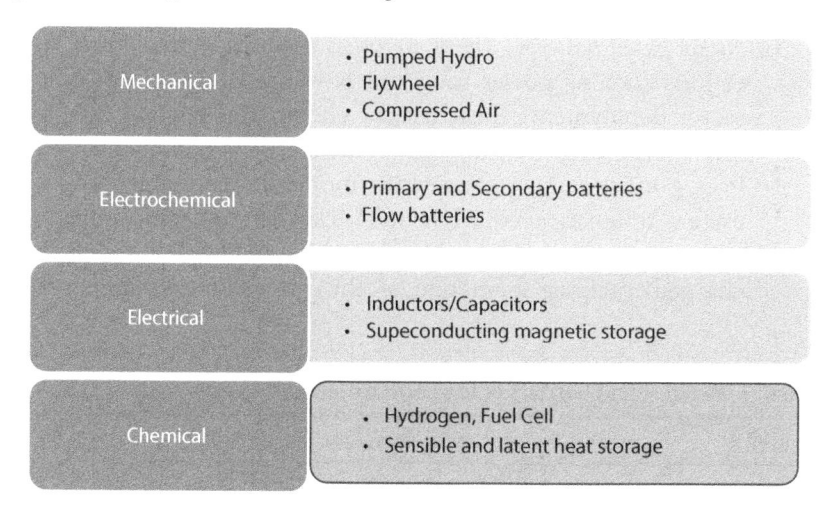

Mechanical	• Pumped Hydro • Flywheel • Compressed Air
Electrochemical	• Primary and Secondary batteries • Flow batteries
Electrical	• Inductors/Capacitors • Superconducting magnetic storage
Chemical	• Hydrogen, Fuel Cell • Sensible and latent heat storage

By definition, those storage devices which convert electrical energy to chemical energy, and vice versa, are termed as electro-chemical storage devices [1]. Electro-chemical storage systems use a series of reversible chemical reactions to store electricity in the form of chemical energy. Batteries are the most common form of electro-chemical storage and have been deployed in power systems in numerous applications. A variety of battery chemistries is available, and they differ in their respective performance parameters. An overview is given below:

(i) *Lead-acid (Pb) batteries:* The lead acid batteries are those secondary cells, having highly porous lead dioxide as positive active material, and finely divided lead is the negative active material. The charging and discharging happen in a dilute aqueous sulphuric acid [2]. The voltage of each cell in a lead acid battery is 2.05 V and, they can generally deliver high value of current. For this reason, lead acid batteries have their deployment in the power sector applications. They have largely been supplanted in Solar photovoltaic-plus-storage systems and residential customer applications by lithium-ion chemistries. A comparison of the benefits of these battery chemistry is given in Table 7.1 [3].

Lead-acid batteries are very safe and highly reliable, which in the past have made them quite popular for backup power supply applications. Based on their application, lead acid batteries can be classified as deep cycle, shallow cycle and sealed maintenance-free batteries.

(ii) *Nickel-based batteries:* These are often used in portable applications such as power tools but have seen limited power sector deployment. These battery chemistries tend to have low energy densities, low power densities, moderate cycle lives, poor round trip efficiencies, moderate costs, and moderate maintenance requirements. Nickel batteries can handle wide ranges of discharging and loading and can be rapidly charged, making them one of the most durable battery

Table 7.1 Benefits and barriers of lead acid batteries.

Benefits	Barriers
• Low cost per watt-hour • Low energy density • High specific power • High discharge currents	• Poor weight to energy ratio. • Relatively poor life cycle. • Slow charging • Transportation restrictions • Frequent watering requirements

chemistries. Finally, some types of nickel batteries (e.g., Nickel-Cadmium) have significant environmental concerns, given their utilization of toxic elements. These batteries also exhibit a memory effect, in which recharging the battery when it is not fully discharged permanently reduces the battery's capacity. The combination of the memory effect, poor power and energy densities, and environmental concerns has limited this chemistry's deployment in power sector applications. A comparison of the benefits of these battery chemistry is given in Table 7.2.

(iii) *Lithium-ion batteries:* In recent years, lithium-ion battery storage becomes the most commonly installed battery technology for power sector applications, deploying in both front-of-the-meter and behind-the-meter applications. Lithium is electropositive, light in weight, non-toxic and comparatively abundant, which makes it a right candidate for many applications. Being highly reactive, use of Lithium in metallic form as cathode material is avoided. Li+ donating compounds like Lithium cobalt oxide, Lithium nickel cobalt aluminium oxide and Lithium nickel manganese cobalt oxide are being generally implemented [4]. Four components in a Lithium-ion battery cell are cathode, anode, electrolyser and separator. Li-ions shuttle from cathode to anode during the charge process, and return shuttle during the discharge process. The shuttle of Li-ions occurs through the electrolyte. Lithium metal oxides are preferred as a cathode material, which predominantly determine the cell properties as they are the Li+ donors. Graphite is widely used as anode material in Li-ion batteries. Electrolyte is a mixture of Lithium salts and organic solvents.

The efficiency and capacity of these chemistries is not as strongly impacted by the depth-of-discharge. Lithium-ion batteries are relatively safe but have the potential for thermal

Table 7.2 Benefits and barriers of nickel based batteries.

Benefits	Barriers
• Comparatively better energy density. • No disposal concerns. • Servicing is easy.	• Limited life. • Limit on discharge currents. • High self discharge.

runaway, in which the internal temperature of the battery rises faster than the heat can be dissipated, which can lead to fires or explosions. Despite safety concerns, lithium-ion has emerged as the primary battery chemistry for behind-the-meter applications, including DPV-plus-storage systems. Careful design and appropriate safety codes can support a safer, but not foolproof, configuration. A comparison of the benefits of this battery chemistry is given in Table 7.3 [5].

(iv) *Sodium-based batteries:* These batteries are primarily deployed for utility-scale applications in the power sector. These batteries are molten metal type battery constructed from sodium and sulphur. The anode is always the sodium, and cathode is either sulphur or Nickel chloride. Beta alumina is the electrolyte used in many cases [6]. These chemistries are made from readily available, inexpensive materials which may improve their economics relative to chemistries with rare element requirements, such as some lithium-ion chemistries. Stringent housing and operating temperature requirements make this chemistry unsuitable for most behind-the-meter applications. A comparison of the benefits of this battery chemistry is given in Table 7.4 [7].

(v) *Flow batteries:* These are relatively new battery chemistry well suited to large-scale and long-duration applications. They are charged and discharged by means of oxidation-reduction reaction of ions of vanadium or similar materials [8]. These batteries have poor energy and power densities, very high cycle lives, low to moderate round trip efficiencies, moderate costs, and relatively low maintenance

Table 7.3 Benefits and barriers of Lithium-ion batteries.

Benefits	Barriers
• High specific energy • High load capabilities • Long cycle and extended shelf life • Maintenance free • High capacity, low internal resistance • Good columbic efficiency	• Protection circuit against thermal runaways. • Degradation at high temperature. • No rapid charge at freezing temperatures.

Table 7.4 Benefits and barriers of sodium-based batteries.

Benefits	Barriers
• High energy densities. • High power densities. • Long cycle life • Low Maintenance • Comparatively lower cost. • Good round trip efficiency	• High temperature operation (above 300°C) • Extra accessory cost for enclosures. • Metallic sodium is combustible when exposed to water.

requirements (primarily for associated pumps). These batteries are very safe and are not prone to fires or other hazards. Furthermore, the power and energy capacities of these batteries are easily and independently sizable, making them a very flexible option from a design standpoint. Given these systems' low energy and power densities and higher relative costs, flow batteries will likely experience limited use in behind-the-meter applications in the near term, where size constraints are more relevant. The major types of vanadium redox batteries are polysulphide bromine battery and Zinc bromine battery.

A detailed comparison of key parameters and performance indicators of various battery technologies are given in Table 7.5 [10].

7.2.2 Energy Storage Landscape and Benefits of Electro-Chemical Storage

Electro-chemical storage or Battery has significant applications in the power sector. Depending on the location of the battery interconnected to the power system (i.e., Behind-the-Meter, Distribution system, or Transmission system), services offered by the battery are different. Value streams for each proposition are listed below:

a) Behind the Meter services:
Behind-the-meter energy storage systems have the capability to act as both generation and load. It represents a potentially unique and disruptive power sector technology capable of providing a range of important services to customers, utilities, and the broader power system. Distribution companies

Table 7.5 Performance indicators of various battery chemistries.

Specifications	Lead acid	Ni-Cd	Ni-MH	Li-on Cobalt	Li-on Manganese	Li-on Phosphate
Specific energy (Wh/kg)	30–50	45–80	62–120	150–250	100–150	90–120
Internal resistance	Very Low	Very Low	Low	Moderate	Low	Very Low
Cycle life (80% DoD)	200–300	1,000	300–500	500–1,000	500–1,000	1,000–2,000
Charge time	8–16 h	1–2h	2–4h	2–4h	1–2h	1–2h
Overcharge tolerance	High	Moderate	Low	Low to trickle charge		
Self-discharge/month (room temp)	5%	20%	30%	<5% Protection circuit consumes 3% month		
Cell Voltage (nominal)	2V	1.2V	1.2V	3.6V	3.7V	3.2–3.3V
Charge cut off voltage(V/cell)	2.40 Float 2.25	Full charge detection by voltage signature		4.20 typical Some go to higher V		3.60
Discharge cut-off voltage (V/cell)	1.75 V	1.00V		2.50-3.00V		2.50V
Peak load current	5C	20C	5C	2C	>30C	>30C
Best result	0.2C	1C	0.5C	<1C	<10C	<10C

(Continued)

Table 7.5 Performance indicators of various battery chemistries. (Continued)

Specifications	Lead acid	Ni-Cd	Ni-MH	Li-on		
				Cobalt	Manganese	Phosphate
Charge temperature	−20 to 50°C (−4 to 122°F)	0 to 45°C (32 to 113°F)		0 to 45°C9 (32 to 113°F)		
Discharge temperature	−2 to 50°C (−4 to 122°F)	−20 to 65°C (−4 to 49°F)		−20 to 60°C (−4 to 140°F)		
Maintenance requirement	3–6 months (toping chg.)	Full discharge every 90 days when in full use		Maintenance-free		
Safety requirements	Thermally stable	Thermally stable, fuse protection		Protection circuit mandatory 11		
In use since	Late 1800s	1950	1990	1991	1996	1999
Toxicity	Very high	Very high	Low	Low		
Coulombic efficiency	~90%	~70% slow charge ~90% fast charge		99%		
Cost	Low	Moderate		High		

have challenges to service the area of their jurisdiction due to congestion, issues of rising demand, generation or transmission unavailability. Batteries are deployed to meet the power outages. BESS have the capability to help retail customers to reduce their electricity bills. BESS are technically capable of grid interactivity, as opposed to previous/existing storage systems, which is primarily designed for back up services only.

Batteries, in addition to the reliability issues, employ behind-the-meter storage to ensure high-quality power, especially in electronics manufacturing or in data centres where power quality issues can significantly impair end-products or services.

Adoption of battery storage helps reduction of consumption from the grid. Bill savings can come both from energy charges (time-of-use) as well as fixed demand charges.

b) Transmission and Distribution Services:

Other applications of batteries in utilities are the demand-side management. Storage resource can manage peak demand, and/or build utility-scale storage facilities to help time-shift demand and provide a range of important ancillary services. Additionally, storage systems could defer/avoid network upgrades, and/or meet otherwise deficit demand. These are at times cost effective as there is cross-subsidizing involved for certain customer classes.

A new innovative market model in India practices that the generator shall supply despatchable renewable power complemented with thermal power, 24×7, keeping at least 85% availability annually and also at least 85% availability during peak hours. Peak hours will be clearly specified four hours in a day. The generator has to offer power such that at least 51% of the annual energy offered corresponds to renewable energy and the balance is offered from thermal sources. The generator combines battery applications for ensuring required minimum annual availability of 85%.

Smart grid technologies (i.e., smart inverters, energy storage, and smart switches) enable safe operations like islanding, which can provide system benefits such as resilience and renewable integration. Micro-grids can serve the larger distribution grid as well as the economic and resiliency needs of third parties.

Micro-grids are on the forefront of the emerging distribution energy ecosystem that challenges the traditional power market. In the future, utilities will continue to forge partnerships with energy storage companies and other vendors who are active in the micro-grid space.

c) Grid Ancillary Services:

Battery energy storage has multifunctional applications in the grid ancillary services. These include, a) Dynamic secondary frequency regulation, b) Primary frequency response c) Stabilize the grid responses with short duration back-up storage. The advantages of the battery energy storage systems, for these applications, includes it can be charged and discharged and can provide bidirectional operation. Lower value of static inaccuracy and higher round trip efficiencies [9].

An overview of the applications of battery energy storage systems in the Utility scale applications is given in Table 7.6

The other direct and indirect operational benefits of the battery energy storage are given in Table 7.7.

Table 7.6 Overview of applications of battery storage in utility applications.

Application	Feature	Duration of service
Arbitrage	Trading of energy, buy at low cost during off peak and sell at high cost during peak time.	Hours
Firm capacity	Provide reliable capacity to meet peak system demand.	Hours
Operative Reserve for primary frequency response	Responses to unpredictable variations in demand and generations.	Seconds
Ramping and load following	Follow longer term changes in electricity demand.	Hours
Transmission and Distribution replacement	Reduced loading on transmission and distribution systems during peak times.	Hours

Table 7.7 Direct and indirect benefits of battery energy storage systems.

Direct operational benefits	Indirect operational benefits
• Contingency reserves. • Power Regulation. • Load following.	• Reduced operating reserve. • Reduced start-up and shut down costs. • Improved heat rate and reduced emissions.

7.2.3 Drivers and Barriers in Implementation of Energy Storage Systems

Battery-based systems are very much portable and could be relocated to new areas when no longer needed in the original location. This can essentially improve the utilization of these systems. These systems can charge during periods of lower demand and meet demand during peak hours storage and could also be directly interconnected to the distribution network to address periods of oversupply of distributed renewable energy systems.

Cost has been the certain barrier for the implementation of these systems. The costing includes both capital expenditures and operational costs which are at times recurring expenses. Although recycling is an option, the cost viability in the recycling procedures presents a barrier to the options for users to choose for recycling of the batteries.

7.3 Design and Performance Parameters for Electro-Chemical Storage

The design of any storage element is decided by the hours of back-up required. However, each cell chemistry being different in charging and discharging patterns, it is essential to have definitions to characterize the battery performances. The key technical characteristics of an electro-chemical storage and its significance is given in Table 7.8 [10].

7.3.1 Design Basis for Large Storage Application

The design of a storage system is at all times customised to the requirements. Although the requirement varies, the general basis of the design is to adopt a system size which optimises the performance and cost. Before discussing the design of the storage system, it is essential to understand the difference between the functional and non-functional requirements in a battery-based storage system.

Table 7.8 Definitions of key battery technical characteristics.

Rated Power Capacity	Total possible instantaneous discharge capability (in kilowatts [kW] or megawatts [MW]) of the battery, or the maximum rate of discharge that the battery storage can achieve, starting from a fully charged state.
Energy Capacity	Maximum amount of stored energy (in kilowatt-hours [kWh] or megawatt-hours [MWh]) a battery can hold (capacity may also be represented as 'Usable Energy Capacity' or 'Operating Energy Capacity' that reflects the highest percentage of the total energy capacity recommended to preserve battery performance).
Energy/Power Density	Measure of the energy or power capacity of a battery relative to its volume (kW/L, kWh/L).
Specific Energy/Power	Measure of the energy or power capacity of a battery relative to its weight (kW/g, kWh/g).
Storage Duration	Amount of time storage can discharge at its rated power capacity before depleting its energy capacity.
Cycle Life/Lifetime	Amount of time or cycles a battery storage system can provide regular charging and discharging before failure or significant degradation.
Round-trip Efficiency	Ratio of the energy charged to the battery to the energy discharged from the battery.
Self-discharge	Reduction of stored energy of the battery (% of charge/time) through internal chemical reactions, rather than through discharging to perform work.

> *Functional requirements* are those which describe a function, for example, it signifies something that a storage system does. The most common function of a battery energy storage system is to import and export the active power stored. Another functional requirement relevant to battery energy storage is the duration of charge. For ex. A 4 MWh storage

can deliver 4 MW of power if discharged in an hour. But, in case, if it discharges 1MW power it can be utilized for 4 hours' duration.
➢ *Non-Functional requirements* are those which are required, but do not signify any function in the energy storage system. These include redundancy requirements, maximum and minimum boundaries of the charging and discharging voltage, local regulatory compliances if any, special conditions regarding protection and permits, recycling options, etc.

The following are, in general, the sequential steps adopted for the designing of an electro-chemical battery-based storage system.

a) Identification of the storage requirements, and performing a pre-assessment for viability check.
b) Initial sizing of the functional requirements based on an energy storage simulation model. This model is essentially derived from thumb rules, and inclusion of the basic loss components.

First value of storage size, in AH = (Storage Requirement in AH / Losses).

a) Perform the iterative sizing, including the time varying functional requirements, and accelerated losses.
b) Convert the design results, to Storage sizing. This is carried out by matching the design value with the rated characteristics of a commercial battery.
c) Apply other non-functional sizing requirements, to fine-tune the sizing.
d) Evaluation of levelized storage cost. The equation for evaluating the levelized storage cost is as follows:

$$\text{Levelized cost of Storage, LCOS} = \frac{\Sigma(C_t + O_t + Ch_t + D_t)*(1+r)^{-t}}{\Sigma MWh_t *(1+r)^{-t}}$$

where,
C_t is the Capital Cost in the year t.
O_t is the Operation and Maintenance cost in the year t
Ch_t is the charging cost in the year t.

Electro-Chemical Battery Energy Storage Systems 243

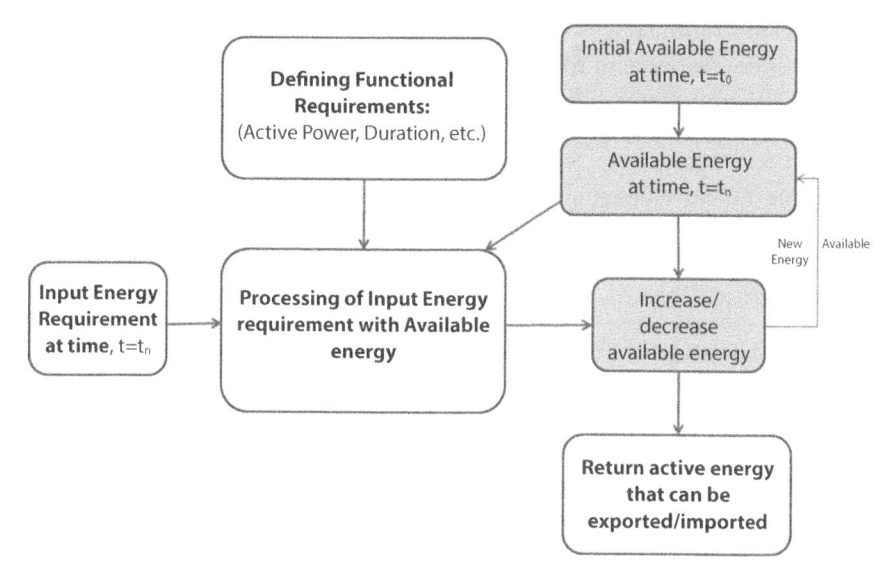

Figure 7.1 Design basis for energy storage evaluation using time dependent simulation model.

D_t is the Decommissioning and waste management cost in the year t.

MWh_t The amount of electricity generated in MWh in the year t.

$(1 + r)^{-t}$ = Discounted factor for the year t, with r being the discount rate.

It is always recommended to use any time dependent simulation models to evaluate the design. This would ensure the match of the system size with the requirements over a duration. A typical flow of the design using simulation model is given in Figure 7.1.

7.4 Case Study From Industry

Storage business is essentially going to emerge as a service industry. The power grid is always under stress, owing to the momentary change of power requirements in the demand side. This forces the grid operators to ramp up or ramp down the power in accordance to the high and low demand period. The energy storage at the demand side is a workable solution to cater the fluctuating power demands. The following is a use case from the food industry, detailing the financial and operational benefits of battery energy storage in the demand side.

The case study from a Solar-Grid Hybrid solution implemented in a small-scale wood industry is furnished in the tables below (Table 7.9a to 7.9d)

Table 7.9a Energy requirements of load.

Connected Load:	108kW
Contract Demand:	90kVA
Average daily usage/hr in Zone 1:	16920kWh@ INR 7.2/kWh
Average daily usage/hr in Zone 2:	6120 kWh@ INR 10.8/kWh
Average daily usage/hr in Zone 3:	10360kWh @ INR 5.4/kWh

Table 7.9b Battery system.

Single Battery rating:	200Ah / 12V @ C10
Energy Storage in Battery:	12 V *200 Ah = 2.4 kWh
Battery Depth of Discharge:	80%
Maximum output from 1 Battery:	2.4*0.8 = 1.92 kWh
Total storage from 60 Battery:	60*1.92 = 115.2 kWh
Nett energy storage after system losses (roundtrip efficiency @ 90%)	103.7 kWh
Nett stored energy used / month	3,111 kWh

Table 7.9c Solar power generation system.

Array (DC) capacity:	100 kWp
Assuming CUF @ 16%, Energy yield / month	100 × 4 kWh × 30 days = 12,000 kWh
Monthly Day time load consumption from utility in Zone 1	16920 kWh − 12000 kWh + 3111 kWh = 8,031 kWh
Monthly load energy requirement in zone 2	6120 kWh
Storage energy utilised/month in zone 2	3,111 kWh
Excess energy drawn from Grid for zone 2	6,120kWh − 3,111 kWh = 3,009 kWh

Table 7.9d Expenditure and savings.

Expenditure for ON Grid Solar Power Plant, 100 kWp capacity with Storage capacity 144 kWh	INR 60.00 Lakhs
Savings in Monthly Energy Charges (Zone 1):	INR 7.20 × 8,889 = INR 64,000
Savings in Monthly Energy Charges (Zone 2):	INR 10.8 × 3,111 = INR 33,599
Net Savings	INR 64,000 + INR 33,599 = INR 97,600
Number of years required for payback	INR 60,00,000/ (INR 97,600*12) = 5.12 years
<u>Summary Remarks:</u> Adoption of battery storage can help the industry to manage the demand side energy requirements, with attractive paybacks.	

7.5 Best Practices in Battery Maintenance

Major checks in battery maintenance include:

a) *Regular inspection and maintenance*
   To ensure optimum performance, a routine inspection in every 30 days is recommended. This maximizes the battery life. The routine inspection may include:
   (i) Battery state of charge.
   (ii) Proper cleaning of the battery from dirt, and keeping it dry.
   (iii) Inspection of terminals, screws, clamps and breakage of cables, loose connections. Terminals to be free from corrosion.
   (iv) To ensure full charge, in case of storing battery for longer duration.
   (v) Topping of the battery, to ensure electrolyte at its desired level.
   (vi) To check for any physical damage in the battery due to heating.

b) *To ensure battery health and safety:*
   The handling of batteries shall be done with utmost care, as there are chances of acid spillover. Also, batteries generate explosive gases and care must be taken to keep flames and sparks away from these. Metallic tools and conductors can cause short circuit and sparks even.
c) *Regular battery testing:*
   A digital battery tester is a preferred option to carry out most of the testing requirements in a battery. Voltmeters, hydrometers, and discharge testers are also used for testing of battery. The instruments and various tests carried out are given in Table 7.10.
d) *Proper battery charging:*
   It is required to ensure the safety precautions, before the charging of the battery. These includes turning off the charger, isolating any flames and sparks, proper ventilation, following manufacturer's instructions, etc. Lead acid batteries are usually charged in three stages, constant current (boost), constant voltage (absorption), and float charge. It is evident to check for any loose connections, electrolyte spilling, or any signs of physical damage before the charging of the battery.

The following are some of the major causes for battery failure:
a) Manufacturing faults
b) Short circuit/Dead cells

Table 7.10 Test instruments used in battery maintenance.

S. no.	Testing instrument	Feature
1	Hydrometer	State of Charge
2	Digital battery tester	Conductance, Resistance, Voltage, determination of early battery failures, etc.
3	Adjustable load tester	To determine the starting capacity of the battery
4	Constant rate discharge testers	To check the capacity of the battery, by discharging at a pre-set current.

c) Wear and Tear
d) Physical damage
e) Sulphation
f) Overcharging or undercharging
g) In correct application
h) Internal break of cells

7.6 End of Life Cycle of Batteries

The demand for a resource-efficient and economically feasible recycling scheme for storage system is the need of the hour. The first life of the electro-chemical batteries are expected to last up to 10 years, depending on the technology. But the batteries can be employed for a second-life application after the recycling process.

The highest utilization (at 85%) of lead is in the production of lead acid batteries [11]. Out of this 60% of lead comes from the recycling. Most of the components in a lead acid battery can be recycled and re-utilized using low-energy input processes [12]. Sequential steps involved in the recycling process of a lead acid battery are shown in Figure 7.2:

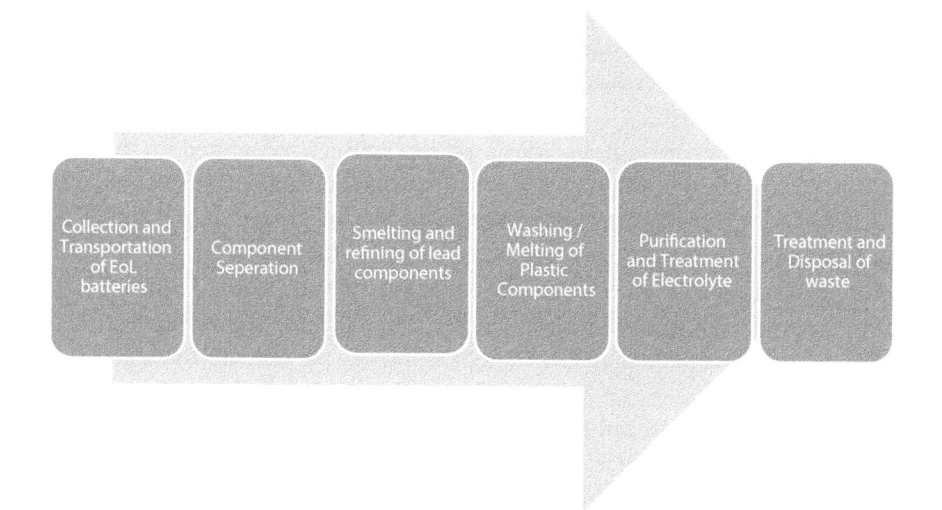

[Flow diagram: Collection and Transportation of EoL batteries → Component Seperation → Smelting and refining of lead components → Washing / Melting of Plastic Components → Purification and Treatment of Electrolyte → Treatment and Disposal of waste]

Figure 7.2 Sequence of steps in recycling of lead acid battery.

7.6.1 Major Recyclable Products from the Process

a) *High Purity refined Lead:* The furnace smelted lead is refined to its purest form and used to produce alloys that can be used to make new lead batteries. Approximately up to 99% of the lead can be recovered, recycled and reprocessed.
b) *Recyclable Plastic Components*: The recyclable plastic components are melted to form plastic pellets, which are then used as other plastic goods including new battery casings. [13].
c) *Electrolyte:* The electrolyte used in the lead acid battery, generally sulphuric acid, is recovered and used as a treating agent for cleaning contaminants. Post-purified sodium sulphate extract can be used as detergent [14].

7.6.2 Disposal Measures

There are various disposal control measures to be adopted during the recycling of lead acid batteries. Major points from the relevant guidelines [15], in this regard are the following:

a) Avoid draining of the batteries at collection point.
b) Availability of well-ventilated and coated, acid resistant concrete ground for storage of lead acid batteries.
c) Use of acid resistant containers for leaking batteries.
d) Used batteries has to be considered in the category of hazardous waste.
e) Keeping upright positions of the battery, separated by a non-conducting material, or in sealed secured containers.
f) Adoption of fully automated and enclosed operations for dismantling and separation of lead acid batteries.
g) Fugitive emissions in the recycling process can be minimized by use of negative pressure enclosures. This should have proper exhaust with high efficiency particulate air filter [16].
h) Ensuring low temperature operations of molten fumes; this essentially reduces the amount of fumes.
i) Requirement of an affluent treatment station to treat all the water used in the recycling process.
j) Safety for the workers, to avoid any exposure to contaminants. The battery industry's voluntary initiative recommends for

workers blood lead concentrations of less than 30 microgram/dL [17].

There are many stakeholders involved in the life cycle management (LCM) of a lead acid battery, like manufacturers, retailers, scrap dealers, secondary smelters and consumers. Improper disposal is a result of the informal approaches in the LCM. Some of the suggestions to formalize the recycling process include the following:

a) Proper refund scheme and mechanism for the consumers. This encourages the collection of used batteries.
b) Manufacturer's commitments in take back policy.
c) Awareness, at large, of the environmental and health hazards in the improper handling of the lead.
d) Developing adequate capacities, trained inspectors, and laboratory facilities.

There are two advantages to the recycling of the lead acid batteries. There is scarcity in the lead resource to meet the scaled-up demand in growing sectors like automobile, energy storage devices, and solar applications. Second, of course, the reduction of lead waste in landfills.

7.7 India Energy Storage Mission

By developing battery manufacturing expertise and scaling its domestic production capacity, India can build economic advantage in this sector. This section highlights the major extracts from the policy measures for making India a hub for battery manufacturing, as proposed by NITI AAYOG [18]. Development of India's battery manufacturing needs to be planned in three different stages, as follows in Table 7.11.

The following are the key challenges to scaling India's battery industry:

a) *Low mineral reserves:* India has a good reserve of Manganese, Aluminum, and Graphite, but does not have reserves for most important Li-ion components like lithium, cobalt, or nickel [19].
b) Early-stage battery manufacturing Industry.
c) *Multiple Stakeholders:* Presence of multiple stakeholders in the Industry, Government, and regulatory bodies keeps the coordination difficult.

Table 7.11 Three-stage development for battery manufacturing in India.

	Stage 1	**Stage 2**	**Stage 3**
Objectives	Creating Environment for battery manufacturing growth	Scaling of supply chain strategies	Building end to end cell and battery manufacturing supply chain
Actions	Incentivize and encourage direct investment in the growth of a battery pack assembly. Developing partnership and consortium for joint research and investment, and battery recycling. Selective choices of individual companies in pursuing battery cell manufacturing.	Identifying and optimizing the cost through effective supply chain strategy.	Facilitating coordination of multiple stake holders.

Table 7.12 Expected economic opportunity from battery manufacturing.

	Cumulative battery requirements (GWh)	**Total market size (INR Lakh crores)**	**Domestic manufacturing**
Stage 1	120	1.3 to 1.4	Battery packs
Stage 2	970	6.1 to 8.9	Battery pack + Limited cell production
Stage 3	2,410	11.7 to 17.1	End to End battery manufacturing

7.8 Conclusion

This chapter highlighted an overview of the various aspects in an electro-chemical storage. These include energy landscape, storage applications, design basis and performance parameters of an electro-chemical storage, a typical use case from an industrial case study, and overview of recycling aspects. The observation from case study concludes that electro-chemical energy storage is a solution for meeting the continuous energy demands, with appreciable payback, and with advancements in the technology the cost would further drop down.

References

1. D. Pavlov, *Lead-Acid Batteries: Science and Technology*, Elsevier, 2011.
2. D.A.J. Rand, P.T. Moseley, in: P.T. Moseley, J. Garche (Eds.), *Energy Storage with Lead-Acid Batteries, in Electrochemical Energy Storage for Renewable Sources and Grid Balancing*, Elsevier, 2015, pp. 201–222.
3. Battery University (2018). "BU-201: How does the Lead Acid Battery Work?," 31 05 2018. [Online]. Available: http://batteryuniversity.com/learn/article/lead_based_batteries.
4. Gianfranco Pistoia, *Lithium-Ion Batteries. Advances and Applications*, 1st Ed., Elsevier, 2014. ISBN: 9780444595133.
5. G. Zubi, R. Dufo-Lopez, M. Carvalho, G. Pasaoglu, The lithium-ion battery; state of the art and future perspectives, *Renew. Sustain. Energy Rev.* 89 (2018) 292–308.
6. Kopera, J.J.C., 2004. Inside the Nickel Metal Hydride Battery. Cobasys, MI, USA. http://www.cobasys.com/pdf/tutorial/inside_nimh_battery_technology.pdf. Accessed 08/11.
7. Yao Wang, Yukun Liu, Yongchang Liu, Qiuyu Shen, Chengcheng Chen, Fangyuan Qiu, Ping Li, Lifang Jiao, Xuanhui Qu, Recent advances in electrospun electrode materials for sodium-ion batteries, *Journal of Energy Chemistry*, Vol. 54, 2021, pp. 225-241, ISSN 2095-4956.
8. Eduardo Sánchez-Díez, Edgar Ventosa, Massimo Guarnieri, Andrea Trovò, Cristina Flox, Rebeca Marcilla, Francesca Soavi, Petr Mazur, Estibaliz Aranzabe, Raquel Ferret, Redox flow batteries: Status and perspective towards sustainable stationary energy storage, *Journal of Power Sources*, Vol. 481, 2021, 228804, ISSN 0378-7753.
9. Motte Cortés, 2018: Agustín Motte Cortés, "Battery Energy Storage Systems providing Frequency Containment Reserve under various regulatory frameworks of selected countries", Master's Thesis, Fraunhofer ISE, 2018.
10. Akhil, Abbas, Georgianne Huff, Aileen Currier, Benjamin Kaun, Dan Rastler, Stella Bingquing Chen, Andrew Cotter, *et al. Electricity Storage Handbook.*

SAND2013-5131. DOE, EPRI, NRECA. July 2013. https:// www.sandia.gov/ess-ssl/lab_pubs/doeepri-electricity-storage-handbook/.
11. H. Ibrahima, B.A. Ilincaa, J. Perronb, Energy storage systems—Characteristics and comparisons, *Renewable and Sustainable Energy Rev.* 12 (5) (2008) 1221–1250.
12. Geoffrey J. May Alistair Davidson Boris Monaho Lead batteries for utility energy storage: A review. *J. Energy Storage.* Vol. 15, February 2018, pp. 145–157.
13. ILA (2015). Lead recycling fact sheet. London: International Lead Association; 2015 (http://www.ila-lead.org/UserFiles/File/ILA9927%20FS_Recycling_V08.pdf, accessed 3 January 2017).
14. UNEP (2003). Technical guidelines for the environmentally sound management of waste lead-acid batteries. Secretariat of the Basel Convention. Basel Convention series/SBC No. 2003/9. Geneva: Basel convention Secretariat; 2003 (http://www. basel.int/Portals/4/Basel%20Convention/docs/pub/tech-guid/tech-wasteacid. pdf, accessed 3 January 2017).
15. Environmentally sound management of spent lead-acid batteries in North America: Technical guidelines. Montreal: Commission for Environmental Cooperation; 2016 (in English, French & Spanish) (http://www3.cec.org/islandora/en/item/11665-environmentally-sound-management-spent-lead-acid-batteriesin-north-america, accessed 25 January 2017).
16. OSHA (2002) Lead – secondary lead smelter. In: eTools [website]. Washington (DC): Occupational Safety and Health Administration; 2002 (https://www.osha.gov/SLTC/etools/leadsmelter/index.html, accessed 3 January 2017).
17. EUROBAT (2013). Battery Associations from North America and Europe, the Middle East and Africa join forces to strengthen workers' protection. Press release, 19 June 2013 (https://eurobat.org/battery-associations-north-america-and-europe-middle-east-and-africa-join-forces-strengthen-workers%E2%80%99, accessed 26 January 2017).
18. India's Energy Storage Mission: A Make-in-India Opportunity for Globally Competitive Battery Manufacturing. NITI Aayog and Rocky Mountain Institute, 2017. http://www.rmi.org/Indias-Energy-Storage-Mission.
19. Elsa A. Olivetti, Gerbrand Ceder, Gabrielle G. Gaustad, Xinkai Fu, Lithium-Ion Battery Supply Chain Considerations: Analysis of Potential Bottlenecks in Critical Metals, *Joule*, Vol. 1, Issue 2, 2017, pp. 229-243.

8

Simulation of Charging and Discharging a Thermal Energy Storage System Involving Phase Change Material

S. Sanyal[1]*, A. Borgohain[2] and S.P. Gupta[2]

[1]*Central University of Jharkhand, Ranchi, India*
[2]*Bhabha Atomic Research Centre, Mumbai, India*

Abstract

This chapter discusses some of the important aspects involved in the design of a thermal energy storage system and presents numerical study and simulation of melting and solidification of a Phase Change Material (PCM) using ANSYS FLUENT. A 3D simulation model of the experimental set-up is developed which consists of a finned u-tube immersed into a PCM enclosed in a cylindrical shell. Scalable meshes for the whole geometry as well as the one-fourth symmetry sector are generated to simulate the melting and solidification of PCM for two cases of without and with fins attached to the fluid pipe. The assumptions, equations involved in numerical modelling, the software-specifications, and distribution of PCM and fin temperatures and liquid fraction of PCM in various cases are discussed. The numerical model developed in this work can be applied to other similar configurations within the domains of properly defined materials and boundary conditions and can be suitably scaled up for developing a large-scale latent thermal storage system.

Keywords: Thermal energy storage, Phase Change Material (PCM), simulation, melting, solidification, numerical model, fins

8.1 Introduction

Energy is the basis of modern society and is important for the survival of humanity as well as the development of civilization. Effective management

---

Corresponding author: shubhamsanyal321@gmail.com

Umakanta Sahoo (ed.) Energy Storage, (253–276) © 2021 Scrivener Publishing LLC

of energy resources is one of the facets of a mature technology that would move the economic status of a country from normal to the heights of societal development. Ever since the 1950s oil peaks and subsequent energy crises, issues of carbon emissions, intolerable pollution levels, global warming, climate change, and energy security have urged the scientific community to advance towards cleaner and sustainable sources of energy. However, most of the renewable energy sources such as solar and wind, have an intermittent nature and there always remains a distinction between the supply and demand of energy about the location and time concerned. Therefore, energy storage technologies are proposed to solve or mitigate this problem. Energy is usually stored in energy storage systems in various forms, namely, mechanical, electrical, electromagnetic, chemical, biological, electrochemical and thermal. These forms of energy storage can be further subdivided depending upon the technology and are listed in Table 8.1. These energy storage technologies are necessary components for judicious and effective utilization of renewable energy sources and energy conservation.

Table 8.1 Various types of energy storage technologies [1].

Form	Technologies
Mechanical	Compressed Air Energy Storage (CAES), fireless locomotive, flywheel energy storage, gravitational, potential energy, hydraulic accumulator, pumped-storage hydroelectricity
Electrical, electromagnetic	capacitor, supercapacitor, superconducting magnetic energy storage
Chemical	biofuel, hydrated salts, hydrogen storage, hydrogen peroxide, power to gas, vanadium pentoxide
Biological	glycogen, starch
Electrochemical	flow battery, rechargeable battery, ultra-battery
Thermal (general)	storage of sensible, latent and thermochemical heat
Thermal (specific)	brick storage heater, cryogenic energy storage, liquid air energy storage, liquid nitrogen engine, eutectic system, ice storage air conditioning, seasonal thermal energy storage, solar pond, steam accumulator

Thermal energy storage (TES) is one such technology that has been a major research topic in recent decades in which thermal energy, in general, can be stored as a change in the internal energy of a material as sensible heat, latent heat, and thermochemical heat, or a combination of these. Thus, thermal energy storage (TES) can be categorized into three types: sensible heat storage (SHS), latent heat storage (LHS) and thermochemical storage (TCS). Sensible heat storage (SHS) systems store energy by temperature changes experienced by the storage medium, which can be either a liquid or a solid material. Latent heat storage (LHS) makes use of the latent heat involved in a phase change, the most common case being solid-liquid transitions. The material which undergoes the phase change is known as Phase Change Material (PCM). Heat is stored during melting of the PCM and is subsequently released during its solidification. Thermochemical heat storage (TCS) is based on reversible chemical reactions in which the charging and discharging steps are conducted by carrying out an endothermic and exothermic reaction, respectively.

The stored thermal energy can be used at a later time for various applications such as heating and power generation. This enables the TES system to balance the demand and supply of energy daily, weekly, and even seasonally. It also helps to reduce peak demand, CO_2 emissions and energy consumption while increasing the overall efficiency of the system. This has been very popular for the conversion and storage of various renewable energy sources, such as solar energy, in the form of thermal energy, which is responsible for significant increase in the share of renewable energy in the energy markets.

Among the three storage methods, SHS and LHS systems have similar operational aspects and are known to compete with each other on economic prospects based on the choice of materials and design of the system. SHS systems usually suffer from limitations related to the large storage size commonly required or the temperature swing created from the addition and extraction of sensible energy. On the other hand, an LHS system offers smaller size of the storage unit for the same amount of energy and exhibits smaller heat losses for the same temperature range. This enables an LHS system to store a large amount of thermal energy (or heat) over a small temperature range, maintaining a fairly constant temperature for a period of time which ultimately enhances the operation for the entire thermal storage system. These features make LHS systems a highly interesting alternative to conventional SHS systems.

8.2 Design of Latent Heat Storage (LHS) System

An LHS system consists of four basic components:

1. a heat storage medium or PCM;
2. a heat transfer fluid (HTF);
3. a heat storage unit or PCM containment structure; and
4. a heat exchange medium between the heat source and sink.

The design-process of an LHS system is a complicated process owing to the consideration of time-varying heat loads and flow conditions. This requires a deep understanding of the heat transfer and the phase change processes involved in the PCM(s), along with sufficient quantitative information on factors such as:

1. the extent to which the thermophysical properties of the PCM affect the storage and recovery processes (especially considering the latent heat of the PCM);
2. the mass of PCM required for the storage unit;
3. the effective time of the phase-change process; and
4. the relative importance of conduction (conjugate heat transfer between the PCM and HTF) and natural convection in the PCM melt.

The whole design-process of a latent heat storage system consists of three stages. The first stage is the identification of the suitable PCM, the second stage involves design of the heat exchanger and the final stage consists of performance evaluation, economic analysis and commercialization of the designed prototype.

8.2.1 Identification of Suitable PCM

The choice for suitable PCM is based on the temperature range of application, and judgement of its thermophysical properties and other characteristics. Latent storage systems using phase change materials (PCMs) are highly attractive due to their high volumetric heat storage capacity, compactness, moderate cost and near constant temperature heat storage/retrieval. LHS materials are extensibly used for heating and cooling purposes in domestic, industrial, textile, spacecrafts, electronic equipment and various other sectors covering a wide range of temperatures from below −40 °C to above 900 °C [2]. Some of these applications of LHS materials have been listed in Table 8.2.

Table 8.2 Applications of PCMs [3].

Application	Temperature range (°C)	
	Minimum	Maximum
Spacecraft electronics protection	−269	130
Thermal protection	−269	130
Adsorption refrigeration	−60	350
Cabin heating and refrigeration	−50	70
Transportation	−50	800
Cold production	−40	−10
Floating and cooling	−40	350
Biomedical applications	−30	22
Food preservation	−30	121
Space heating and cooling of buildings	18	28
Solar clergy	20	565
Solar energy storage	20	150
Electronic devices thermal protection	25	45
Heating and cooling of water	29	80
Battery and electronic protection	30	80
Industrial waste heat recovery	30	1600
Desalination	40	120
Exhaust heat recovery	55	800
Industry	60	260
Solar cooling	60	250
Absorption refrigeration	80	230
Microprocessor chips thermal protection	85	120
Solar power plants	250	565

Table 8.3 Suitable PCMs for storage of latent heat [4].

Material	Melting temperature (°C)	Melting enthalpy (MJ/m^3)	Remarks
Water-salt solutions	−100–0	200–300	• Good energy storage density • Eutectic compositions can be used to avoid phase segregation
Water	0	330	• Vapour pressure and corrosion issues become prominent above 100 °C
Clathrates	−50–0	200–300	• Low number of thermal cycles
Paraffins	−20–100	150–250	• Good storage density with respect to mass • Other organic PCMs do not show sub-cooling • Suitable temperature range for most applications • Flammability matter of concern • Usually compatible with metals, but not with plastics
Salt hydrates	−20–80	200–600	• Higher phase change enthalpy than organic PCM • Thermal conductivity similar to water • Disadvantages include phase segregation, sub-cooling, and corrosion
Sugar alcohols	20–450	20–450	• Energy storage density is quite high • Can oxidise in presence of oxygen, need to be used in inert atmospheres

(Continued)

Table 8.3 Suitable PCMs for storage of latent heat [4]. (*Continued*)

Material	Melting temperature (°C)	Melting enthalpy (MJ/m³)	Remarks
Nitrates	120–300	200–700	• Main issues to be considered include vapour pressure, sub-cooling, corrosion, segregation, changes in composition and microstructure, changes in thermal properties, and undesired reactions
Hydroxides	150–400	500–700	
Chlorides	350–750	550–800	
Carbonates	400–800	600–1000	
Fluorides	700–900	above 1000	

Selection of a PCM for a particular application requires evaluation of several criteria based on its material properties as well as characteristics of the overall system (Table 8.3). For use in a latent heat storage system, a PCM should fulfil the following requirements:

1. A melting temperature in the desired temperature range of operation or application;
2. A high value of latent heat of fusion per unit volume to ensure compactness of the storage component;
3. A high value specific heat for additional sensible heat storage;
4. Thermal conductivity should be high in both solid and liquid phases for fast heat transfer;
5. Changes in the volume due to phase transformation should be small, that is, there should be little expansion and contraction of the PCM after melting and solidification;
6. Low degree of supercooling to maintain the same melting/solidification temperature and avoid heat release problems, and a high nucleation rate;
7. The PCM should undergo congruent melting for a constant storage capacity with each freezing/melting cycle;
8. It should be stable after performing a large number of charge-discharge thermal cycles; and
9. It should be less reactive and non-toxic, compatible with the container materials, retaining long-term chemical stability.

Table 8.4 Various methods and instruments used for measurement of thermophysical properties of PCMs [5].

Property	Method/Instrument
Latent heat	DSC, T-History method
Thermal conductivity	T-History method
Thermal diffusivity	Conductimeter
Viscosity	Viscometer, Rheometer
Density	Buoyancy technique
Thermal expansion coefficient	Mechanical dilatometry, optical methods, diffraction techniques

The different methods or instruments used to study the thermophysical properties of PCM are listed in Table 8.4. The thermal stability of the PCM can be determined by performing a large number of charge-discharge cycles followed by measurement of its thermophysical properties. PCM corrosion analysis should also be studied to judge its compatibility with the constituent material of the containment structure.

8.2.2 Design of Heat Exchanger

Once the thermophysical properties, thermal reliability and corrosion analysis of the PCM are studied and the properties and parameters of the component materials are verified, the selection and design of a suitable heat exchanger is done on the basis of the overall heat transfer coefficient on the hot and cold side, the mean log temperature difference, and the effectiveness-NTU method. The designed parameters are simulated and iterated numerically by changing the parameters to achieve a highly efficient heat exchanger geometry. Developing and scaling up of unoptimized prototypes can lead to higher inventory of materials increasing the weight of the system and unnecessary additional costs [6]. Therefore, a detailed optimization study is carried out by developing a numerical tool to optimize the geometric configuration prior to performance evaluation of the LHS prototype. The finally fabricated heat exchanger is filled with the suitable PCM and integrated with a heat source at the charging side and a load on the discharging side.

8.2.3 Performance Evaluation

Before commercialization of the final design, feasibility of the overall system is judged through evaluation of the performance parameters such as energy, exergy and economy. There should be suitable heat transfer between the HTF and the storage media. The system should be long-lasting, non-polluting with low CO_2 footprint and cost-effective with low manufacturing energy and small payback period. With regard to its components, they should be cheap and abundantly available. There should be suitable integration of the components into the facility with proper operation strategy. The final design is compared to others using standard testing procedures. ASHRAE 94-77 and NBSIR 74-634 are two such standard procedures available for testing of thermal storage devices in which the heat loss for the whole charging or discharging process is calculated by integrating the heat flow rates for various time-intervals of the flowing heat transfer fluid (air or water) [7]. During the payback period, there should be easy recycling and treatment, and use of by-products.

8.3 Analysis of Phase Change Systems

Melting and solidification are heat transfer phenomena associated with solid-liquid phase change of a heat storage medium in which the absorption or release of thermal energy is in the active zone occurs via conduction or convection. The essential and common feature of systems undergoing heat transfer due to the solid-liquid phase change is the existence of two different types of boundaries – a static interface, that separates two regions with different thermophysical properties, and a moving interface, that separates the two phases in which the thermal energy is absorbed or released. To carry out heat-transfer analysis of phase change process it is necessary to determine the manner and rate at which the solid-liquid interface moves with time [8], Due to the motion of the solid-liquid interface, the mathematical problems posed are non-linear and only a few exact solutions are available, that too for one-dimensional geometries only, and not for the finite ones.

A PCM can have a discrete phase change temperature with a sharply defined interface, or can experience a phase change over a temperature range in which there is a two-phase region between the solid and the liquid phase, depending upon whether it is a pure substance or a mixture or a commercial-grade material. Presence of undesired cavities and unstable convective motions within the PCM melt due to generation of buoyancy forces caused by density differences and temperature variations during heat transfer further complicates the problem. The heat transfer problem that arises in the presence of phase-change, that is, in the presence of a moving boundary, is known in the literature as the Stefan problem, after the man who studied the thickness of polar ice in 1891. Based on the choice of dependent variable, two types of models, namely, temperature-based models and enthalpy-based models, are used to analyse phase-change heat transfer problems and can be solved using either analytical methods or numerical methods. The equation of energy balance at the interface can be written as:

$$\rho_s \Delta h_m v_{\Sigma} = \left(k\frac{\partial T}{\partial n}\right)_s - \left(k\frac{\partial T}{\partial n}\right)_l \qquad (8.1)$$

where,
  T = Temperature of the solid or liquid;
  ρ = Density of the solid or liquid;
  k = Thermal conductivity of the solid or liquid;
  Δh_m = latent heat of fusion of the material;
  $\vec{v}_{\Sigma}$ = velocity of the interface (normal to itself); and
  $\vec{n}$ = direction vector (normal to the boundary).

The subscripts s and l denote the solid and liquid phases respectively.

Earlier, the temperature-based approach was mostly used in which the energy conservation equations are written separately for solid and liquid regions (either dimensional or non-dimensional forms), with temperature being the only dependent variable. However, this method suffered from a problem with formulation of the interface using finite difference or finite element methods. To simplify the problem, the enthalpy-based approach is used in which the interface is excluded from consideration of calculations and the problem is made equivalent to heat conduction without any phase-change. The mathematical modelling of the LHS prototypes is, thus, quite a complex process. The basic configurations used in numerical analysis involve rectangular, spherical, and cylindrical geometries. Advanced configurations are used for slurries, porous media and surfaces attached to multi-dimensional fins.

The major drawback of PCMs is their low thermal conductivity which results in the slow transient response of the thermal storage system [9]. It has been found that employing finned tubes with different configurations, suitable PCM encapsulation methods and addition of particles with higher thermal conductivity than the PCM, enhance the effective thermal conductivity of the thermal storage system, thereby reducing the effective time required for melting and solidification of the PCM [10]. Other issues include incongruent melting or phase separation, subcooling during solidification, insufficient long-term stability and compatibility issues with storage containers [11]. The subcooling problem can be addressed by use of appropriate nucleating agents, Cold Finger Technique, and increased surface roughness or use of electrolytically polished and porous surfaces. Long-term stability can be ensured by choosing PCM that can withstand more than 1,000 charge-discharge cycles. Corrosion above 700 °C can be prevented by use of containers made of Inconel alloys.

8.4 Simulation

A variety of computational simulation and modelling techniques have recently been developed for analysing the thermal energy storage performance of PCMs. The approaches available are Enthalpy models, Temperature Transforming Model (TTM), Variable viscosity of the medium (VVM), Resistance-Capacitance (RC) models, Conduction Transfer Function (CTF) and Stefan models. The range of accuracy, computational-time and simulation-cost varies for each approach. Examples of sophisticated numerical packages are ANSYS FLUENT [12], COMSOL Multi-Physics [13], and HEATING which mainly use FEM (finite element method) and FDM (finite difference method).

The simulation of the present heat storage unit has been done using ANSYS FLUENT 16.0. To solve flow problems in which melting and solidification take place at one temperature or over a temperature range, ANSYS FLUENT uses an enthalpy-porosity formulation in which the solid-liquid mushy zone is treated as a porous zone with a porosity that corresponds to the liquid fraction of the material. As the PCM melts the liquid fraction (or porosity) of the mushy zone increases from 0 to 1 and decreases from 1 to 0 as the material solidifies. A pressure-based solver is used to model the fluid flow with an approximation of incompressible flow of the heat transfer fluid. Appropriate terms added to the momentum equations account for the pressure drop caused by the presence of solid material which can act as a momentum sink or heat sink or both, whereas turbulence equations account for the reduced porosity in the solid regions.

8.4.1 Equations Involved

The energy equation for enthalpy-based formulation of the melting (or solidification) of PCM takes the form:

$$\frac{\partial(\rho H)}{\partial t} + \nabla \cdot \rho \vec{v} H = \nabla \cdot (k \nabla T) + S \qquad (8.2)$$

where,
- H = total enthalpy
- ρ = density
- $\vec{v}$ = fluid velocity
- S = source term

The total enthalpy (H) is the sum of enthalpy due to reference temperature, sensible heat and latent heat:

$$H = h_{ref} + \int_{T_{ref}}^{T} c_p \, dT + \beta L \qquad (8.3)$$

where,
- h_{ref} = reference enthalpy
- T_{ref} = reference temperature
- c_p = specific heat at constant pressure
- β = liquid fraction or melt fraction
- L = latent heat of the material

The liquid fraction (β) can be defined as:

$$\begin{aligned} \beta &= 0 & &\text{if } T < T_{solidus} \\ \beta &= 1 & &\text{if } T > T_{liquidus} \\ \beta &= \frac{T - T_{solidus}}{T_{liquidus} - T_{solidus}} & &\text{if } T_{solidus} < T < T_{liquidus} \end{aligned} \qquad (8.4)$$

The solution for temperature is an iteration between the energy equation [eq. (8.2)] and the liquid fraction equation [eq. (8.4)]. For proper convergence, an expression for updating the liquid fraction suggested by Voller and Swaminathan [14] is used.

The source term S is added to account for momentum sink due to reduced porosity in the mushy zone and turbulence production due to presence of solid matter in the mushy and the solidified zone. The expression for the source term is given by [15]:

$$S = \frac{(1-\beta)^2}{(\beta^3 + \varepsilon)} A_{mush} (\vec{v} + \phi) \tag{8.5}$$

where β is the liquid volume fraction, ε is a small number (0.001) to prevent division by zero, A_{mush} is the mushy zone constant (10^5 in this case), $\vec{v}$ is the fluid velocity, ϕ represents the turbulence quantity being solved such as specific dissipation rate, turbulent intensity, turbulent viscosity ratio, etc.

8.4.2 Modelling

The experimental set-up consists of a cylindrical shell made of stainless steel with a flange on the top in which holes can be drilled at equal spacing in the pattern of a regular hexagon for insertion of HTF tubes as shown in Figure 8.1. Fins can be attached to the tubes through welding or brazing depending upon the choice of fin material.

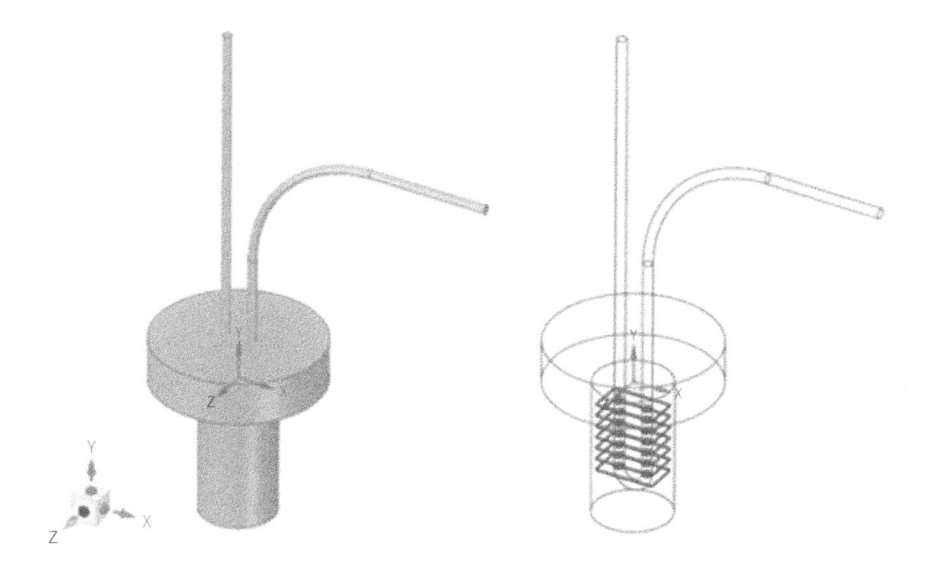

Figure 8.1 Experimental set-up: opaque view (left), transparent view (right).

The simulation model (shown in Figure 8.2) consists of the following components:

(i) A cylindrical vessel made of stainless steel with inner diameter of 90 mm, height of 200 mm and thickness of 5 mm, the top surface having two holes with diameter 11.25 mm with their centres spaced 33.9 mm apart.
(ii) A u-tube made of aluminium with inner diameter of 10.25 mm and thickness of 1 mm.
(iii) 7 fins with dimensions [50 mm*50 mm*2 mm] made of aluminium which are attached to the u-tube with vertical spacing of 20 mm along the tube.

The dimensions of all the three components along with their constituent material are specified in Table 8.5.

The air filled in the u-tube acting as the HTF forms the first fluid zone and the PCM between the inner surface of the cylinder and outer surface of the fin-pipe coupled system forms the second fluid zone. The properties of the PCM are given in Table 8.6.

The meshing of the whole geometry was done in ANSYS. The meshes for full body and one-fourth symmetry sector have been shown in Figure 8.3 and Figure 8.4 respectively. To perform a much simpler analysis and check scalability of the mesh, one-fourth symmetry sector of the set-up was also modelled.

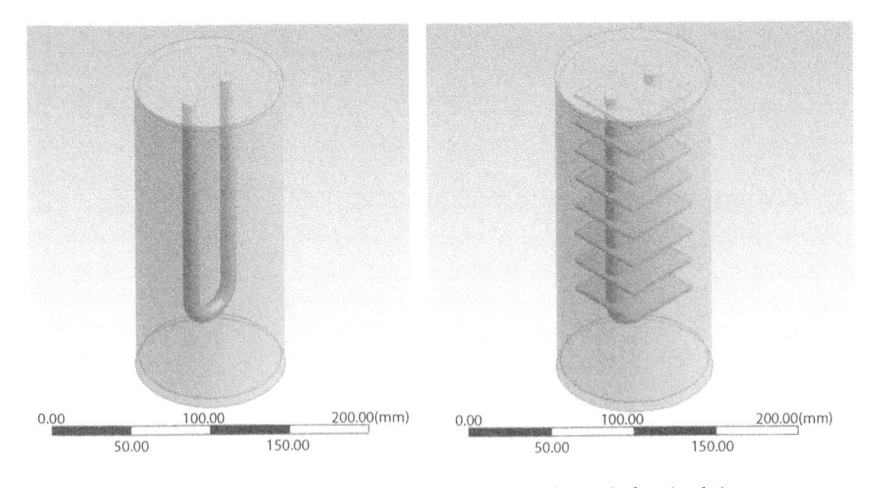

Figure 8.2 Simplified simulation model: without fins (left), with fins (right).

Table 8.5 Dimensions and constituent material of components.

Component	Dimension (in mm)		Material
Cylindrical vessel	Inner diameter	90	Stainless Steel
	Height	200	
	Thickness	5	
U-tube (180° bend)	Inner diameter	10.25	Aluminium
	Thickness	1	
	Bend radius	16.95	
Fins	Length	50	Aluminium
	Height	50	
	Thickness	2	
	Spacing	20	

Table 8.6 Properties of PCM.

Melting point	415 K
Melting Enthalpy	1,10,000 J/kg
Specific Heat	1170 J/kg (solid)
	1730 J/kg (liquid)
Thermal Conductivity	0.72 W/m-k (solid)
	0.57 W/m-k (liquid)
Density	2006 kg/m^3
Viscosity	0.02 kg/m-sec
Coefficient of thermal expansion	3.629×10^{-4}/K

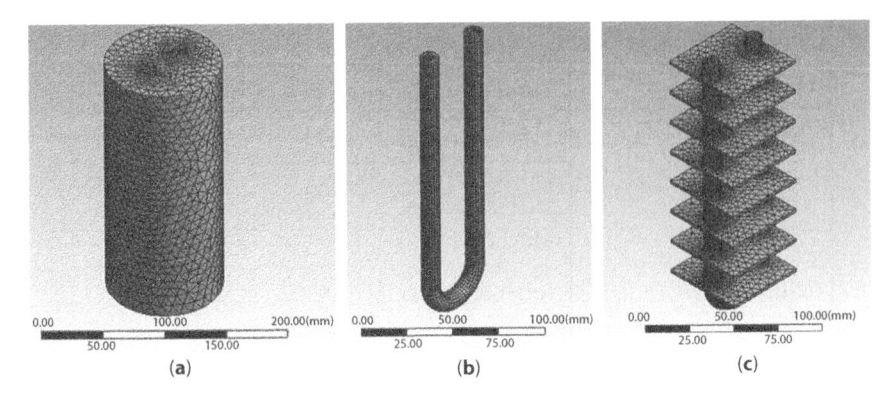

Figure 8.3 Mesh of full body – (a) Main component (PCM), (b) pipe, (c) pipe-fins coupled system.

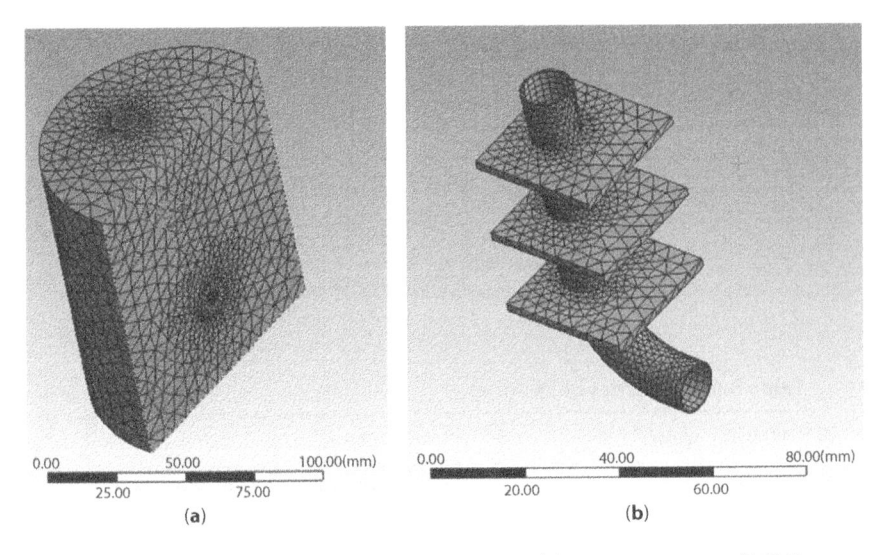

Figure 8.4 Mesh of the one-fourth symmetry sector – (a) Main component (PCM), (b) pipe-fins coupled system.

The one-fourth symmetry was obtained by first dividing the whole geometry into 2 halves by a symmetric horizontal plane (cutting both cylinder and u-tube diametrically) which was further cut by a symmetric vertical plane (converting the 180° bend of the u-tube to a 90° bend tube). The 1-4th symmetry sector alone consists of 3,47,244 cell elements of size 2mm and 78,252 nodes.

8.4.3 Transient Analysis

The SIMPLE algorithm was used with pressure-velocity coupling scheme for both steady and transient formulations. Gravity was turned on with a value of -9.81 m/s^2 along the vertical direction and Bounisque approximation was used for formulating density of the PCM with thermal expansion coefficient of 3.629*10^{-4} per K. Piecewise linear profiles were used for specific heat and thermal conductivity whereas other properties including viscosity were assumed constant. The flow was modelled as viscous with turbulent intensity and turbulent viscosity ratio limited to 5% and 10 respectively. For melting, the boundary conditions applied were:

1. Zero heat flux at outer surface, that is, insulated outer surface;
2. Inlet air velocity of 10 m/s at 430 K and 1,01,325 pascal operating pressure; and
3. Pressure outlet at operating pressure of 1,01,325 pascal with total backflow temperature of 430 K.

For solidification, the temperatures of flowing air and that of total backflow were changed from 430 K to 298 K (ambient temperature), with all other parameters remaining the same.

To check the absolute convergence of the flow and energy residuals and ensure convergence of the solution, in accordance with Navier–Stokes equation, Continuity equation and Energy equation, steady state analysis of the set-up was run before the transient analysis. Graphs of mass average static temperature of the PCM and fins and mass average melting fraction of the PCM were plotted and recorded at time step. With initial absolute criteria of 10^{-3} for the flow residuals and 10^{-6} for the energy residual being set, the steady-state analysis was run to check the convergence of the solution. The solution was seen to converge well meeting the required convergence criteria.

8.5 Results and Discussion

8.5.1 Scalability of Mesh

To check the actual scalability, the simulations were carried out with meshes generated for both full as well as 1-4th symmetry sector geometries. In each case, the domain extents for the mesh was seen to range from -45.0 mm to +45.0 mm in the x-direction, -200 mm to 0 mm in the y-direction and from -45.0 mm to +45.0 mm in the z- direction, with scaling factor of 1000 along the three directions. During simulation, the variation of PCM temperature and melting fraction were found to be exactly the same. Hence, the prototype can be suitably scaled up. Moreover, the symmetry factor can save considerable computational time for numerical simulations. To avoid repetition, only the simulation results for full geometry are presented here and discussed in the upcoming sections.

8.5.2 Melting

The simulation of melting and solidification of PCM were carried out for two cases -

Case (i) - without fins, and
Case (ii) - with fins.

The results are discussed in the following sub-sections:
Figure 8.5 depicts the variation of temperature and liquid fraction during melting of PCM. The melting of PCM was simulated by turning on the flow of hot air at 430 K at t=0 seconds. In case (i), after the application of heat at t=0 seconds, the melting of PCM begins at 4.4 seconds. It takes 14.7 hours for the PCM completely. The total heat absorbed by the system is 186.89 kJ, out of which the melting of the PCM absorbs 50.5 KJ. In case (ii), the melting of PCM begins at 0.9 seconds and it takes 10.6 hours for the whole PCM to melt completely. Out of 139.68 kJ of total heat supplied to the system, 37.29 kJ is absorbed during melting of PCM. Thus, addition of fins fastens the melting process by approximately 28%. It also reduces the absorption of heat flux required for melting the same amount of PCM by 26.17%. The melting of PCM, in case of fins involved, is depicted through counters of static PCM

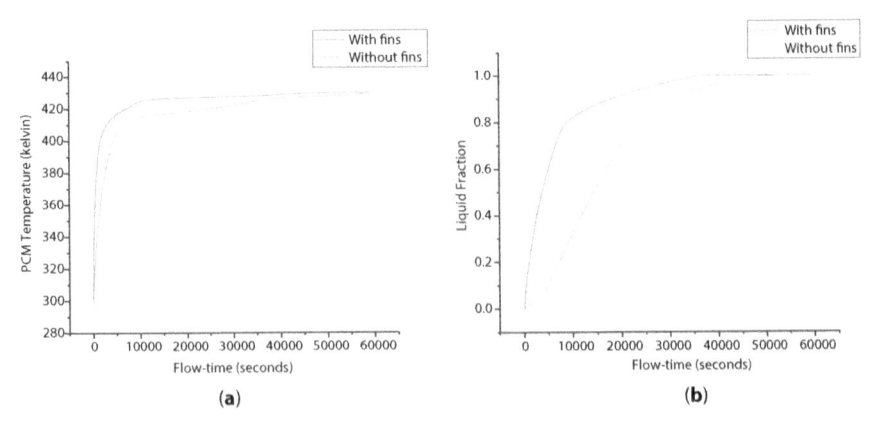

Figure 8.5 Graphs showing variation of temperature and liquid fraction during melting of PCM.

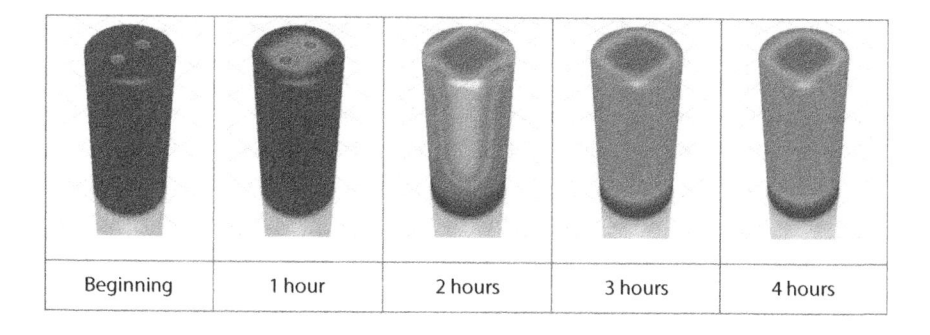

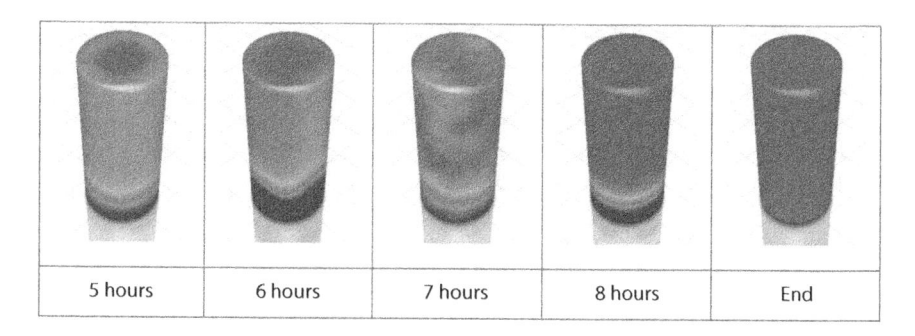

Figure 8.6 Contours of static temperature showing melting of PCM (in case of fins involved).

temperature in Figure 8.6 with interval of 1 hour. The BGR (Blue-Green-Red) colour scheme is used to represent the temperature scale from 298 K (minimum, represented by blue in the beginning) to 430 K (maximum, represented by red in the end).

The liquid fraction counters in Figure 8.7 represent the propagation of melt-front or molten PCM layer through the mushy zone from inwards, near the heating element, to outwards, near the adiabatic outer surface of the cylinder with passage of time. The BGR (Blue-Green-Red) colour scheme is used to represent the liquid fraction scale from 0 (fully solid, represented by blue in the beginning) to 1 (fully liquid, represented by red in the end).

8.5.3 Solidification

After the complete melting of the PCM, the solidification was simulated by turning off the flow of hot air. The cooling process is assumed to start at

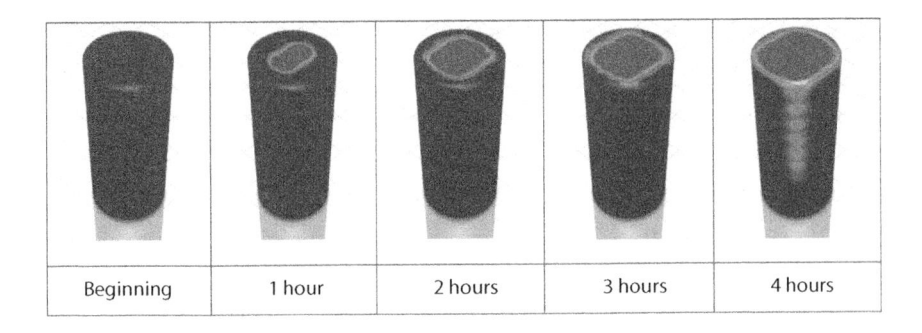

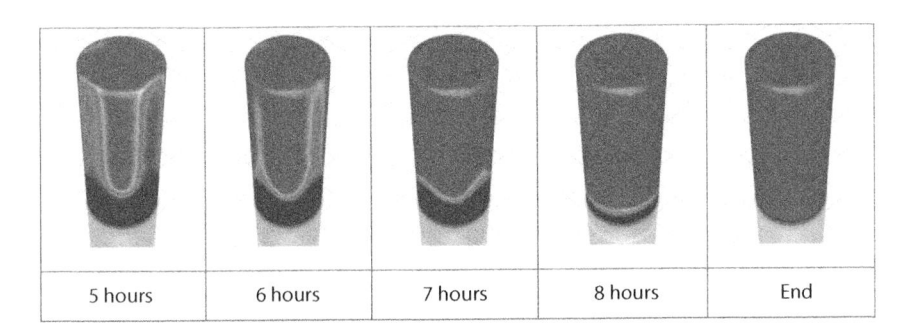

| Beginning | 1 hour | 2 hours | 3 hours | 4 hours |

| 5 hours | 6 hours | 7 hours | 8 hours | End |

Liquid Fraction: 0.00 — 0.10 — 0.20 — 0.30 — 0.40 — 0.50 — 0.60 — 0.70 — 0.80 — 0.90 — 1.00

Figure 8.7 Contours of liquid fraction showing melting of PCM (in case of fins involved).

t=0 when the PCM is at 430 K and temperature of the heating element just reaches 298 K. The solidification, in both the cases, begin at 0.1 seconds. In case (i), the solidification completes in 2 hours with net release of 174 kJ of energy, whereas in case (ii), it takes 1.62 hours with net release 74 kJ of energy.

It can be observed from Figure 8.8 that solidification, being a conduction-dominated process takes much less time than melting. The addition of fins reduces the solidification time by 19% and increases the net heat flux (or energy) release rate by 57.54%. The temperature contours obtained during solidification of PCM were obviously similar to that of melting with the order being reversed with different time-scale and need not be represented here. Therefore, it can be clearly observed that the addition of fins reduces the absorption of heat flux (or energy) in case of melting and increases the releases of heat flux (or energy) in case of solidification. This

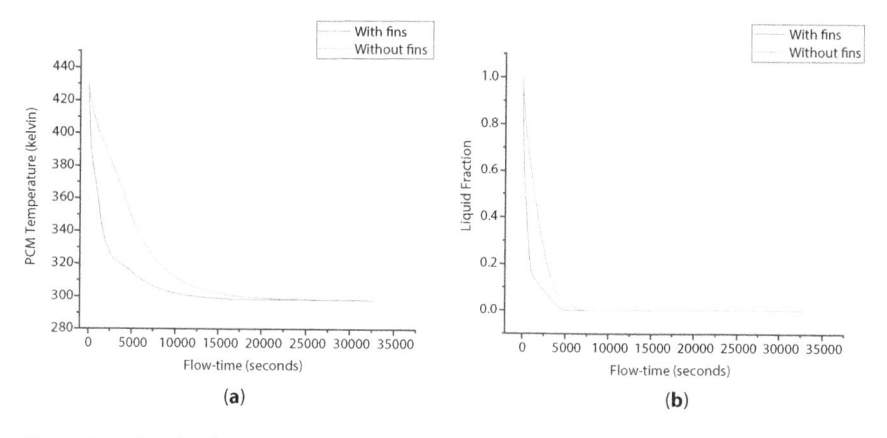

Figure 8.8 Graphs showing variation of temperature and liquid fraction during solidification of PCM.

ultimately reduces consumption of energy while charging and discharging the system.

8.5.4 Performance

In the present case melting and solidification actually are seen to occur over a range of temperatures, without much reversibility of the phase change processes (enthalpy hysteresis). Theoretically, there should have been three distinct regions, namely, the first sensible heating region from 298 K to 415 K, latent heating at 415 K, and finally the second sensible heating region from 415 K to 430 K. The addition of fins increases the surface area by nearly 74%. The total heat (considering both latent and sensible heats), that would have been observed in this case, would add up to be 676.09 kJ and 654.27 kJ in case of without fins and with fins, respectively. Considering the case of actual melting of PCM, without taking into account the temperature swing which occurs in the second sensible heating after significant amount of time followed by complete melting, the total heat supplied in case of without fins and with fins are found to be 186 kJ (11.3%) and 139 kJ (27.64%) respectively. In other words, actual amount of heat absorbed during melting of PCM is found to be 27.64% and 11.3% of the theoretical maximum in case of without fins and with fins, respectively. Similarly, the actual amount of heat released during solidification of PCM is found to be 28% and 37.2% of the theoretical maximum in case of without fins and with fins, respectively. This deviation from the "ideal system" can be accounted for by the local energy non-equilibrium caused due to high heat flux from air to PCM and low heat diffusion due to small thermal conductivity of PCM material in the liquid phase. The actual heat distribution across the cross section of the material is somewhat non-uniform as natural convection begins due to the volume expansion and temperature differences within the liquid phase. This effect is more pronounced in case of fins involved.

8.6 Conclusion

In the present work, some important design aspects of thermal storage systems involving PCMs were discussed and simulation of a thermal storage unit was done using the FLUENT 16.0 package of ANSYS. ANSYS FLUENT uses an enthalpy-porosity formulation, in which, rather than implicitly tracking the dynamic solid-liquid interface, the solid-liquid mushy zone is treated as a porous zone with porosity corresponding to the PCM liquid fraction. A simulation model of the experimental set-up was made which is basically a finned u-tube immersed into a PCM enclosed in a cylindrical shell. The simulation of melting and solidification of PCM was done through flow of hot air at 430 K and ambient air at 298 K through aluminium pipe immersed into the PCM. Two cases of without and with fins of aluminium attached to the pipe were considered. The range of domain extents and variations of temperature, PCM liquid fraction and heat flux are similar for the meshes of both full geometry as well as the one-fourth symmetry sector, which indicates that the model prototype can be suitably scaled up with scaling factor of 1000. It was observed that the addition of fins significantly reduces the time required for phase change, by 28% and 19% in case of melting and solidification, respectively, which can ultimately reduce the consumption of energy while charging and discharging the thermal storage system by up to 26.17% and 57.54%, respectively.

Acknowledgement

The authors would like to extend their sincere gratitude to Bhabha Atomic research Centre (B.A.R.C.) (Government of India), Mumbai, Maharashtra, India and Central University of Jharkhand (Government of India), Ranchi, Jharkhand, India, for the help and support provided for this work.

Abbreviation

TES Thermal Energy Storage
SHS Sensible Heat Storage
LHS Latent Heat Storage
PCM Phase Change Material
HTF Heat Transfer Fluid

References

1. https://en.wikipedia.org/wiki/Energy_storage
2. A. Sharma, V.V. Tyagi, C.R. Chen, D. Buddhi, Review on thermal energy storage with phase change materials and applications, *Renew. Sustain. Energy Rev.* 13, 318–345, 2009.
3. Jouhara, Hussam; Zabnienska, Alina; Khordehgah, Navid; Ahmad, Darem, and Lipinski, Tom. Latent Thermal Energy Storage Technologies and Applications: A Review. *International Journal of Thermofluids*, 5, 2020.
4. Mehling, H., Cabeza, L.F. *Heat and Cold Storage with PCM: An Up to Date. Introduction into Basics and Applications.* Springer, Berlin Heidelberg, 2008.
5. Andrea Frazzica, Luisa F.Cabeza. Recent Advancements in Materials and Systems for Thermal Energy Storage, An Introduction to Experimental Characterization Methods. *Green Energy and Technology*, Springer Nature Switzerland AG, 2019.
6. Hakeem Niyas, Sunku Prasad, P. Muthukumar. Performance investigation of a lab-scale latent heat storage prototype – Numerical results. *Energy Conversion and Management (Vol.-135)*, 188–199, ISSN 0196-8904, 2017.
7. H. P. Garg, S. C. Mullick, A. K. Bhargava., *Solar Thermal Energy Storage*, Springer Netherlands, 1985.
8. Viskanta R., Phase-change heat transfer. In: Lane, G.A. (Ed.), *Solar Heat Storage: Latent Heat Materials, Volume I: Background and Scientific Principles*, 5, 1983.
9. Agyenim, F., Hewitt, N., Eames, P., Smyth, M., A review of materials, heat transfer and phase change problem formulation for latent heat thermal energy storage systems (LHTESS). *Renew. Sustain. Energy Rev.* 14, 615–625, 2010a.
10. Agyenim, F., Eames, P., Smyth, M, Heat transfer enhancement in medium temperature thermal energy storage system using a multitube heat transfer array, *Renew. Energy*, 35, 198–207, 2010b.
11. Frank Bruno, Martin Belusko, Ming Liu, N.H. Steven Tay, Advances in Thermal Energy Storage Systems: Methods and Applications. Woodhead Publishing Series in Energy (Author – Luisa F. Cabeza), 9, 221–268, 2020.
12. https://www.ansys.com/products/fluids/ansys-fluent

13. https://www.comsol.co.in/comsol-multiphysics
14. Voller, V.R., Swaminathan, C.R., General source-based method for solidification phase change. *Numerical Heat Transfer B* 19, 175, 1991.
15. https://www.afs.enea.it/project/neptunius/docs/fluent/html/th/node354.htm

Index

Absorption cycle, 172
Active single-tank thermocline, 20
Active topology, 97
Active two tank system, 9
Active two-tank direct, 9
Active two-tank indirect, 10
Actuators, 52
Adsorption cycles, 173
Aperture area of reflector, 59
Aquifer, 158
Asymmetric supercapacitor, 130

Battery storage system, 90

Cast-iron core crucible design, 56
Central receiver plants, 6
Chalcogenides, 135
Chemical energy storage, 89
Composite supercapacitor, 129
Compressed air energy storage, 89
Corrosion problem in TES-CSP system, 26
Costing of 1 MW Solar PV, 41
CSP projects, 3
CSP receiver, 4
CSP technology, 2
CSP with thermal storage project, 11

Dish system, 7
DNI radiation, 70

EDLC materials, 131
Electrical double-layer capacitor, 126

Electrical energy storage, 84
Electro-chemical storage, 231
Electrolytes, 138
End of life cycle of batteries, 247
Energy storage devices, 83
Energy storage systems, 40
Energy storage technology, 85

Flat plate collector, 179
Flexible parabola, 51
Flywheel energy storage, 87
Fresnel lens, 54
Fuel cell, 90

Global supercapacitor market, 205
Global wind installation, 82

Heat loss and radiation, 62
Heat transfer unit, 54
Hybrid configuration, 95
Hybrid energy storage system, 93
Hybrid supercapacitors, 211

India energy storage mission, 249

Latent heat storage, 24, 163
Lead storage battery, 214
Lead-acid battery, 91
Linear fresnel reflector systems, 5
Lithium-ion battery, 91

Macro and microencapsulation, 165
Materials for latent heat TES systems, 25

277

Mechanical energy storage, 87
Metal oxides/hydroxides, 133
Microsupercapacitors, 210

Nickel cadmium, 91, 213
Nitrides and phosphides, 136

Oscillation damping, 106

Packed-bed storage system, 21
Parabolic solar reflector, 49
Parabolic trough system, 4
Paraffins, 258
Passive thermal storage system, 22
Passive topology, 95
PID logic, 67
Polysulfide bromine batteries, 92
Power fluctuation mitigation, 104
Power installed capacity of India, 39
Pseudocapacitive materials, 132
Pseudocapacitor, 128
Pumped hydroelectric storage, 88

Receiver coil, 53

Scalability of mesh, 269
Sensible energy storage, 22
Separators, 139
Smart supercapacitor, 211

Sodium nickel chloride, 91
Sodium sulphur, 91
Sodium sulphur battery, 215
Solar air heater, 183
Solar fuel, 90
Solar pond, 178
Solar PV/T, 181
Solar tracker unit, 54
Spinning reserve, 107
Status of supercapacitor in India, 125
Super capacitor, 86
Superconducting magnet energy storage, 86
Superconducting magnetic energy storage, 85

Thermal energy storage system, 40
Thermochemical energy storage, 25, 168
Transient analysis, 269

Vanadium redox flow battery, 92

Water based storage, 153
Worldwide CSP plants, 4
Worldwide CSP receiver, 6
Worldwide CSP storage plant, 9

Zinc bromine batteries, 92

Also of Interest

Check out these other related titles from Scrivener Publishing

Other Books in the Series, "Advances in Renewable Energy"

Hybrid Renewable Energy Systems, edited by Umakanta Sahoo, ISBN 9781119555575. Edited and written by some of the world's top experts in renewable energy, this is the most comprehensive and in-depth volume on hybrid renewable energy systems available, a must-have for any engineer, scientist, or student. *NOW AVAILABLE!*

Progress in Solar Energy Technology and Applications, edited by Umakanta Sahoo, ISBN 9781119555605. This first volume in the new groundbreaking series, Advances in Renewable Energy, covers the latest concepts, trends, techniques, processes, and materials in solar energy, focusing on the state-of-the-art for the field and written by a group of world-renowned experts. *NOW AVAILABLE!*

Also by the Same Editor

A Polygeneration Process Concept for Hybrid Solar and Biomass Power Plants: Simulation, Modeling, and Optimization, by Umakanta Sahoo, ISBN 9781119536093. This is the most comprehensive and in-depth study of the theory and practical applications of a new and groundbreaking method for the energy industry to "go green" with renewable and alternative energy sources. *NOW AVAILABLE!*

Other Related Titles

Energy Storage 2nd Edition, by Ralph Zito and Haleh Ardibili, ISBN 9781119083597. A revision of the groundbreaking study of methods for storing energy on a massive scale to be used in wind, solar, and other renewable energy systems. *NOW AVAILABLE!*

Nuclear Power: Policies, Practices, and the Future, by Darryl Siemer, ISBN 9781119657781. Written from an engineer's perspective, this is a treatise on the state of nuclear power today, its benefits, and its future, focusing on both policy and technological issues. *NOW AVAILABLE!*

Zero-Waste Engineering 2nd Edition: A New Era of Sustainable Technology Development, by M. M. Kahn and M. R. Islam, ISBN 9781119184898. This book outlines how to develop zero-waste engineering following natural pathways that are truly sustainable using methods that have been developed for sustainability, such as solar air conditioning, natural desalination, green building, chemical-free biofuel, fuel cells, scientifically renewable energy, and new mathematical and economic models. *NOW AVAILABLE!*

Sustainable Energy Pricing, by Gary Zatzman, ISBN 9780470901632. In this controversial new volume, the author explores a new science of energy pricing and how it can be done in a way that is sustainable for the world's economy and environment. *NOW AVAILABLE!*

Sustainable Resource Development, by Gary Zatzman, ISBN 9781118290392. Taking a new, fresh look at how the energy industry and we, as a planet, are developing our energy resources, this book looks at what is right and wrong about energy resource development. This book aids engineers and scientists in achieving a true sustainability in this field, both from an economic and environmental perspective. *NOW AVAILABLE!*

The *Greening of Petroleum Operations*, by M. R. Islam *et al.*, ISBN 9780470625903. The state of the art in petroleum operations, from a "green" perspective. *NOW AVAILABLE!*

Emergency Response Management for Offshore Oil Spills, by Nicholas P. Cheremisinoff, PhD, and Anton Davletshin, ISBN 9780470927120. The first book to examine the Deepwater Horizon disaster and offer processes for safety and environmental protection. *NOW AVAILABLE!*

Biogas Production, Edited by Ackmez Mudhoo, ISBN 9781118062852. This volume covers the most cutting-edge pretreatment processes being used and studied today for the production of biogas during anaerobic digestion processes using different feedstocks, in the most efficient and economical methods possible. *NOW AVAILABLE!*

Bioremediation and Sustainability: Research and Applications, Edited by Romeela Mohee and Ackmez Mudhoo, ISBN 9781118062845. Bioremediation and Sustainability is an up-to-date and comprehensive treatment of research and applications for some of the most important low-cost, "green," emerging technologies in chemical and environmental engineering. *NOW AVAILABLE!*

Green Chemistry and Environmental Remediation, Edited by Rashmi Sanghi and Vandana Singh, ISBN 9780470943083. Presents high quality research papers as well as in depth review articles on the new emerging green face of multidimensional environmental chemistry. *NOW AVAILABLE!*

Bioremediation of Petroleum and Petroleum Products, by James Speight and Karuna Arjoon, ISBN 9780470938492. With petroleum-related spills, explosions, and health issues in the headlines almost every day, the issue of remediation of petroleum and petroleum products is taking on increasing importance, for the survival of our environment, our planet, and our future. This book is the first of its kind to explore this difficult issue from an engineering and scientific point of view and offer solutions and reasonable courses of action. *NOW AVAILABLE!*